Genetic Engineering

edited by
David Paterson

British Broadcasting Corporation

575·1

PAT

59661

Published by the
British Broadcasting Corporation
35 Marylebone High Street
London W1

SBN 563 08451 0

First published 1969
© British Broadcasting Corporation and contributors 1969

Printed in England by Eyre and Spottiswoode Limited
The Thanet Press Margate

9001560157

Genetic Engineering

Edited by David Paterson

The future of man is likely to depend in part on our ability to manipulate the inheritance of living things. The fundamentals of this ability have come to us through molecular biology, but so far most of molecular biology has been concerned with micro-organisms —bacteria and viruses—rather than with the possibility of controlling the heredity of the higher animals, or changing the design of man.

Even though we are not in sight of these goals, some of the issues raised by genetics research make even today's genetic engineering an extremely important subject. In these talks, based on a BBC Third Programme series produced in 1968 by David Paterson, eight speakers give an account of the most significant developments in their specialised fields— evolution and heredity; breeding animals for food; reproduction control; chromosomes in wheat; the provision of genetic advice; population genetics; and micro-organisms. In the final talk the distinguished American geneticist, T. M. Sonneborn, discusses the conflicts raised by genetic engineering with current standards of conduct, and the choices that are open to society.

Necessarily, most of the speakers did some crystal ball gazing—and in the time lag between broadcast and publication one speaker at least has been shown to be right. In the article on reproduction control, Robert Edwards, the Cambridge University physiologist who hit the news headlines so recently, anticipated success in the routine fertilisation of human eggs and foresaw some of the moral issues raised by this research.

Contents

1 Evolution, heredity, and eugenics
James F. Crow   7

2 Breeding for food
Alan Robertson   17

3 Reproduction: chance and choice
R. G. Edwards   25

4 The chromosome construction and
reconstruction of wheat
Ralph Riley   33

5 Genetic counselling
A Physician   41

6 Genes and populations
Anthony Allison   49

7 Microbes and molecules
D. J. Cove   58

8 Man's choices
T. M. Sonneborn   65

Introduction

Genetic Engineering is a new technology and over the next few decades, there can be little doubt that the future of man will depend heavily on his ability to manipulate the inheritance of living things. The fundamentals of this ability are with us now – the ultra-rapid growth of molecular biology in the last fifteen years has provided us with an astounding insight into the mechanisms of inheritance and how we can alter these mechanisms at will.

So far, much of molecular biology has been concerned with micro-organisms – bacteria and viruses – but the nature of scientific research dictates that some day we will become *au fait* with the molecular events that control the heredity of the higher animals, even man himself. And given this information, we will then probably be equipped to make drastic changes in the design of man.

Even though that very distant prospect may seem daunting, genetic engineering has a great deal to offer us at present. In this series of eight talks, I asked each of the speakers to give an account of the most significant developments in his specialised field and to examine the social and moral problems that could arise through the application of these developments. The final talk, given by the distinguished American geneticist, T. M. Sonneborn, ends the series on a note of confidence that probably we are not speeding to the biological doom predicted so confidently by many commentators.

David Paterson

1 Evolution, heredity, and eugenics

James F. Crow

Genetics has attracted an uncommon amount of interest in recent years. I pick up a popular weekly and find jokes and cartoons about DNA. James Watson's highly personal account of the discovery of the chemical structure of DNA was a recent best-seller at the bookstores, along with the latest novel about adultery in suburbia. Some months ago the then President of the United States announced at a press conference that biologically active nucleic acid had been made in the chemical laboratory. Recently the newspapers carried articles about human cytogenetics, stimulated by the report that a mass murderer was found to have an extra Y chromosome.

What opportunities has genetics brought? What problems, both scientific and social, has it raised? These are some of the questions that this series of talks will attempt to answer.

Narrowly, genetics is the study of heredity – of biological inheritance. It seeks to understand why corn plants are more like other corn plants than they are like cabbages or kings. It asks why related individuals are alike in many ways and yet so different.

The first half of the twentieth century was also the first half century of genetics as a recognised science. During this period the major beneficiary of genetic knowledge was agriculture. Selection, inbreeding, hybridisation, and artificial fertilisation are now routine procedures in animal breeding. In addition, the plant breeder has many more techniques at his disposal – mainly because plants can reproduce in a much larger variety of ways than can animals. As a result of applied genetics, high-yielding, disease-resistant crop varieties are commonplace and plants and livestock are made to order for particular environments and for specialised human needs.

This will undoubtedly continue, but it is likely that the major practical emphasis of genetics in its second half-century will shift more and more to man himself. As man understands his own heredity, he will find new ways of curing, ameliorating, or preventing genetic diseases. Human genetics will become a matter of greater concern because, as more and more infectious diseases are reduced to insignificance, genetic diseases become relatively more common.

Modern genetics has had diverse origins. One of these is Charles Darwin and the theory of evolution by natural selection. At first Darwinism was rejected out of hand by a great many people, even biologists. The idea that man is related to the birds and flowers and – for some reason, worst of all, to the apes – was regarded as absurd. Then came grudging acceptance epitomised by the Victorian lady who hoped that the theory was wrong but, if it turned out to be true, hoped that it would never become widely known. Now, of course, it is fully accepted *and* widely known. With the realisation that all life has evolved comes the realisation that we, too, are evolving and that we can, if we choose, influence the process.

Darwin did not understand the nature of heredity, for it was not known to biologists at that time. Only one man knew the rules, and hardly anyone knew him. The story of Gregor Mendel's experiments on garden peas, how he discovered the laws of inheritance and how these principles went almost completely unknown until their rediscovery at the turn of the century has been told many times.

Mendel showed that the observed facts of heredity can be understood by postulating hereditary particles – now called genes – that are transmitted from generation to generation. Within a few years after the discovery in 1900, it was demonstrated beyond any doubt that the genes are carried by chromosomes – the tiny bodies in the cell that microscopists had observed for many years. Moreover the genes are arranged in single file along the chromosome and their exact position can often be determined.

Thus, by World War I, geneticists knew the rules of inheritance and the location of the genes. But there was al-

8

most no insight into the physical and chemical nature of the gene. This had to wait for two more developments. One was the discovery – in 1946 – that bacteria and viruses have Mendelian inheritance or, in plainer language, they have sex. This means that breeding experiments are possible. These tiny organisms have turned out to be far more willing to reveal their inner nature to the curious biologist than the higher plants and animals have been. Partly this is because they are simpler; but a more important reason is that bacteria and viruses multiply so rapidly and can be grown in such immense numbers that events with probabilities as small as one in a hundred million or less can be studied quantitatively.

At the same time as the micro-organisms were being exploited so effectively, chemists were learning how to work with large molecules. It became clear that the gene is fundamentally deoxyribonucleic acid, or DNA as it is abbreviated. It also became clear that the key to an understanding of the gene lay in the structure of DNA. The double helical DNA model, as proposed by Watson and Crick in 1953, is now familiar to every school biology student, if not to his parents.

The model of the atom, as conceived by Niels Bohr, was a landmark in the physical sciences. By its very structure the model offered explanations for a number of previously mysterious physical and chemical problems and suggested experiments to solve others. The Watson-Crick model of DNA has had the same influence in biology. A glance at the double coil structure immediately suggests explanations for some of the most fundamental and hitherto mysterious properties of the gene. This is highlighted by Watson and Crick's famous understatement in their original paper: 'It has not escaped our notice that the specific pairing we have postulated immediately suggests a possible copying mechanism for the genetic material.'

The process of self-copying, the nature of mutation, and the manner of information storage all are suggested immediately by the double spiral structure. The information that is transmitted from parent to child is encoded in the sequence of the four kinds of bases that bridge the double helix ten times in every full·twist and hundreds of times in every gene. The reason for the exquisitely perfect self-mimicry of the genetic material is apparent in the complementary double structure, in which each half copies its opposite half every time the molecule divides. Finally, the process of mutation – the characteristically biological property of copying mistakes rather than what was there before the mistake was made – finds a ready explanation in the way the zipper mechanism faithfully copies whatever is there, right or wrong – much as errors in ancient manuscripts were perpetuated as they were transcribed.

The structure of DNA has been fruitful in suggesting further experiments by which it has been possible to understand how proteins are made. Furthermore, the nature of the instructions for protein synthesis is known – the code by which the gene transmits its message as to which proteins to make.

A most remarkable property of the genetic code is its universality. With possibly minor exceptions, a sequence of DNA is translated into protein the same way in a virus, in bacteria, in higher plants, in higher animals, and in ourselves. Since it is utterly implausible that the code would have been independently invented several times, this must imply that the same coding system has been used for a very long time – back to the time when all of these diverse forms of life could be traced to a common ancestor. I guess it is fair to say that the genetic code is probably the oldest thing that the geneticist knows anything about.

These extremely deep insights came very quickly. The whole science of molecular biology is about fifteen years old. It is paradoxical that the problem that to biologists had once seemed to be the most difficult of all – the nature of the gene – is now so deeply understood. I find it indeed wonderful, perhaps a bit strange, that we can understand the inheritance of something when we have no fundamental understanding of the something itself. We understand in a very deep way the principle by which muscular dystrophy is inherited; yet we do not understand in any similar degree at all what causes the muscles to degenerate. Perhaps even more striking is the fact that we understand the basis for the inheritance of the ability to think and to remember, even though we do not have much idea of what either thought or memory really is.

Such deep insights are certain to lead to practical results. One is a better understanding, and therefore a better control, of virus infections. Cellular and molecular understanding is gradually leading to ways to decrease the genetic barriers to tissue and organ transplants.

Diseases caused by defective genes can be treated by repair processes that are closer to the gene and its immediate products. More and more, it will be possible to replace empirical methods with more satisfactory treatments based on fundamental understanding.

Another consequence of a deep knowledge of heredity is the realisation that our complex environment may harbour genetic hazards. Geneticists have known since the 1920's that radiation increases the rate of production of harmful mutations. Fortunately this discovery was made before the wide use of nuclear energy so that the appropriate precautions could be taken. We now realise that a number of chemicals may be mutation-producing. It is important that those to which man may be exposed be tested for mutagenicity before any wide use. Fortunately, by use of tests systems involving micro-organisms and cell cultures, this can be done with considerable precision and without exorbitant costs.

Since a great deal of human misery is caused by new mutations, we might hope that one day ways will be found to reduce the spontaneous mutation rate. That this is not hopeless is indicated by the fact that some chemicals are known to do this in bacteria. If you are worried about the possibility that – if the mutation rate is lowered – we may seriously deplete the supply of genes for future evolution, consider the wide extent of genetic variability in the population now. We have pigmies and giants, we have a wide range of shapes and colours, we have talents such as Shakespeare, Newton, Michelangelo, and Mozart. The existing pool of genes carries an enormous diversity although – as J. B. S. Haldane once pointed out – if we ever want a race of angels we shall probably have to have new mutations, for the present supply of genes includes neither those for the wings nor for the requisite moral perfection.

We cannot discuss evolution and heredity without realising that evolution in general, and human evolution in particular, is still going on. We are forced to the realisation that man's evolutionary future depends on choices made by individuals and by society, wittingly or unwittingly.

To decide that we should do nothing about human evolution does not mean that man will not determine its course. It means instead that choices made on other grounds – our economic decisions, our tax policies, our educational system, the availability of transportation, medical practices – become determinants of our genetic future.

Shortly after Darwin's theory was proposed, his cousin, Francis Galton, started thinking about the possibilities for human biological improvement. He noted that natural selection is a slow, inefficient, erratic process. Its means are death and sterility. Its only criterion of excellence is the capacity to survive and reproduce. It has no foresight, no systematic way of passing a road block by detouring, no way of pursuing other than the most immediate and elementary aims.

Galton suggested that by birth selection all these difficulties might be overcome and progress could be made toward desirable goals. Evolution under human influence could have foresight, could have broadly humanitarian aims, and could be accomplished by much more humane means than natural selection. It was with these idealistic motives that eugenics began. Its controversial subsequent history is well known.

Should man try to influence his genetic future ? Have our very low death rates led to the virtual elimination of natural selection so that our population is slowly deteriorating from accumulation of mutations? Is the present pattern of differential birth rates good or bad? Obviously there are no easy answers.

Fortunately, genetic changes are slow. Whatever is happening is measured in generations, so we can afford to take time for a careful consideration. Genetic problems are certainly not of the urgency of, for example, the problems of unrestricted growth of the total population or the tragedy of nuclear war – the twin problems of overpopulation or no population.

Nevertheless, there is little doubt that much human misery is caused by deleterious genes. Some of this could certainly be prevented by genetic means, if society wishes to do so. Compulsory eugenics programmes have few supporters, but there is considerable discussion of what steps might be taken voluntarily by individuals.

I'll consider some areas where genetic knowledge is being used and some others that may become possible in the future.

The first is Genetic Counselling. The geneticist can often specify quite exactly the probability of abnormal or diseased children from pedigree information. In some cases this prognosis depends on simple genetic knowledge. For example, parents who produce a child with cystic fibrosis can be told that the recurrence risk is one in four. Likewise a woman who has a brother and uncle with haemophilia can expect half her sons to be affected.

Two areas of rapid advance in recent years are especially worthy of mention. One is the detection, usually by chemical means, of persons who carry harmful recessive genes but do not manifest the disease. This is possible for many of the abnormal haemoglobin types and for various inborn errors of metabolism. Thus, two normal persons who carry the same hidden gene could learn in advance of the risk of an affected child.

The second area of rapid advance is in human chromosome analysis. Although most instances of multiple congenital anomalies caused by chromosome unbalance are sporadic, some are from high risk families. The usual explanation of such families is that one parent has a chromosome rearrangement, which predisposes him to have children with an abnormal chromosome composition. Such rearrangements can often be detected by microscopic examination of white blood cells, so persons who have a high risk of abnormal children can be identified.

It is now possible to obtain embryonic cells from the fluid surrounding the foetus. Such cells can be cultured and studied for chromosome irregularities. If these are found it can be predicted with certainty that the child will have multiple anomalies and be mentally retarded. This tragedy could be avoided by therapeutic abortion.

At present a couple with a chromosome rearrangement often avoids having children entirely. Therapeutic abortion of all abnormal embryos would permit them to have children with no greater risk than other parents. There are similar possibilities for embryologic detection of other types of genetic disease. Also technical advances are likely to permit detection at an earlier prenatal age. That the possibility of aborting abnormal embryos raises religious, moral, and ethical questions goes without saying. It is another instance where a technological advance may force a revision of conventional western views.

Artificial insemination is already used voluntarily when the husband is sterile or has a severe inherited disease. Sperm may be refrigerated and stored indefinitely. The donor is typically anonymous, being known only to the doctor.

The great geneticist, H. J. Muller, advocated a greater freedom of choice for the married couple in choosing a sperm donor. They could then choose a person who possesses those traits they would most like to see in their children. Muller urged especially the use of sperm from a person no longer living in order that traits throughout his lifetime could be considered; for example, one could avoid donors who developed mental or physical disease late in life or were prematurely senile.

There is no doubt that the population could change appreciably if this were widely practised and if there were any consistent pattern in the choice of donors. The effect would not be as great as one might naively expect, however. This is because of what Francis Galton called regression. Children tend to be somewhat closer to the population average than their parents. Furthermore, the Mendelian mechanism leads to wide fluctuations from the average in specific instances.

Although egg transfer has not, to my knowledge, been tried in humans, it is now routine in some mouse laboratories and has been done in cattle. There appear to be only technical difficulties to human egg transplant. If there were a strong incentive to do so, these difficulties could probably be overcome and the same opportunities made available when the wife is sterile or carries a genetic disease.

Vegetative or clonal reproduction is now possible in frogs. This is done by transferring the nucleus of a body cell from one frog into the egg of another, the egg's original nucleus having been previously removed. In some cases a normal frog developed; it was then a genetic carbon copy of the one that donated the cell nucleus.

The possibility of vegetative reproduction in man has been discussed by Haldane, by Muller, and most recently by Joshua Lederberg. It may still be some time in the future, but something that can be done in one vertebrate cannot be counted out as a possibility for others.

The knowledge explosion in molecular biology is certain to have its practical consequences. We can hope eventually to intervene at all stages in gene action. The idea is not new; for example, we have been using insulin treatment in diabetes for some time. What is new is the possibility for intervention at levels closer to the gene and its immediate products, thereby making the repair more specific and more complete.

Microbial genetics has a host of techniques for removal, addition or replacement of genes and these will presumably one day be applicable to man. They raise the hope for eventually replacing a particular defective gene without altering the rest of the heredity. I should emphasise, though, that microbial techniques that are so powerful in producing pinpoint genetic changes involve means for selecting those which are desired from many more unwanted types. This, of course, would not be acceptable for mankind. The use of molecular engineering to change heredity may be some time away.

I expect that society will be much more willing to accept molecular intervention than such things as vegetative reproduction, since the first applications of molecular engineering will involve the individual rather than his descendants.

Whether man wants to look to his future and consider changes other than the reduction in incidence of serious diseases is more controversial. Are we content merely to keep man from getting worse genetically? Is there enough agreement on both means and ends for any kind of positive eugenics programme? There does not seem to be any wide agreement at present. There are both biological and social uncertainties. I am once again reminded of Haldane who suggested that if the unscrupulous get rich and the poor have more children, at any rate we should expect some moral improvement!

Eugenics has had a checkered history. It started out with high idealism. Later on it got mixed up with various dubious and even pernicious political movements; the much greater possibilities, for both good and bad, that are beginning to appear call for much less simplistic assumptions and more sophisticated consideration of all aspects of human biology. We might prefer to avoid a disquieting discussion of the extent to which man should try to influence his genetic future; it is always easier to adopt the comfortable expedient of ignoring the subject. Yet, it is certain that inventive biologists and chemists are going to find more and more effective ways of influencing human evolution. That society will use them wisely is much less certain.

2 Breeding for food

Alan Robertson

Today most of our domestic animals, with the exception of some highly selected kinds of poultry, have been bred essentially without benefit of Mendel. Only in the last twenty years or so has genetics really made any impact on animal breeding, and in some kinds of animals the results are still not very obvious.

Animal improvement started about ten thousand years ago. Midden sites in the Near East and on the Black Sea coast which can be dated at about that time show a high proportion of bones of domesticated sheep – the dog was also domesticated about the same time and cattle some two thousand years later. The last of the large mammals was the horse, around 3000 BC. Domestic poultry appeared in India at about the same time and the last of our domestic birds to appear in Western Europe was, of course, the turkey, native to Central America. Cortez mentions seeing white peacocks in Mexico which were probably turkeys. These were introduced to Europe at the beginning of the sixteenth century and spread very rapidly, incidentally causing a great deal of etymological confusion. The English name implies that they come from Turkey, the French that they come from India, and even the zoologists confuse the issue by using for them the Latin word for the guinea fowl. An interesting by-product of modern genetics is that some of this pre-history of domestication can be confirmed by studying blood groups and other such differences in our present animals.

The present pattern of breeds began to take shape only two hundred years ago. Each small region had then its own local breed, often distinguished by special colour patterns. Around 1800, enthusiastic farmers began to improve their local cattle, mainly by establishing standards of practical merit and breeding the best to the best. In these hands, such breeds as Herefords, Aberdeen Angus, and Ayrshires have their first recognisable origins. These improved local breeds then gradually replaced their inferior neighbours, a process still going on today, but on an international scale. Black and white Friesian cows, coming immediately from Holland, but perhaps originally from West Denmark, now dominate our milk production and in the past five years Charolais beef cattle from France have made their mark. But British breeders would be justifiably angry if I forgot to mention the great impact of the reverse process. British breeds of cattle, sheep and pigs played a large part in animal production throughout the world. In terms of social history, though, I wonder how much of this was due only to the merits of the animals themselves and how much as well to the rapid expansion of British trade going on at the same time. Cattle as well as trade can follow the flag.

The animals available when Mendel's work was rediscovered in 1900 were undoubtedly better for their own specialised purposes than the animals of a hundred years earlier. But at the same time, animal breeding had been formalised (I am tempted to use the word fossilised) and given a social basis. The early improvers had practical economic aims foremost in their minds. Their animals had as well certain visible trademarks, often in colour and external appearance. Unfortunately some of these trademarks begin to achieve merit in their own right, so that in the eyes of breeders and farmers alike they almost supplanted more practical values. In any case it is often very difficult for the farmer to decide whether an animal does well because of its genes, or because of his superior management.

For almost thirty years, the rediscovery of Mendelism had little or no effect on animal improvement. In the first enthusiasm over the new discoveries which made clear how to make *de novo* kinds of animals or plants with specific, simply inherited characteristics, several enthusiasts set out to make a new breed by combining, for instance, the high milk yield of the Friesian cow with the high fat content of the Jersey's milk. Such hopes quickly vanished, because milk yield and fat content are not inherited as simply as the colour of hair or whether or not the cow has horns. We speak of the former kind of characteristic as showing continuous variation, probably controlled by many genes, and much influenced by the treatment that the animal receives as well. The necessary theory for the application of the new ideas in practice was not worked out completely almost until 1950.

Although genetics is now important in both plant and animal improvement, the ideas current amongst practical breeders of the two have very little in common. The plant breeder can cultivate plants by the million, and an individual grain of wheat is worth commercially a mere 1/5000th of a penny. Because, in most species, plants are self-pollinating it is easy to make genetic copies of the same individual and so to evaluate different alternative varieties very accurately. But the breeder of the dairy cow, for instance, deals with a genetic unit which is a distinct and unrepeatable individual worth about £100 in the market, and therefore in the hands of individual breeders in tens of animals rather than in millions.

Because of the social aspects of animal breeding, the application of genetics raises sociological as well as scientific questions. As geneticists, we often think we know what to do but lack the chance to try. Opportunities have been offered in recent years by the spread of new techniques both of husbandry and of breeding. The development of intensive animal production, especially of poultry, has meant the emergence of large producers who must, to stay in business, be sure that they work with the best genetic stocks available. It is easy for them to make good economic judgements of different breeders' material, and this has meant an alteration of the breeders' viewpoint away from the traditional trademark type of breeding towards selection for economic characteristics. Indeed, to the old poultry breeder, modern kinds of poultry are indescribable. In other words, professionalism amongst producers has led to professionalism among breeders. The same tendency can be seen in pig production and to a smaller extent in sheep. In dairy cattle the main new factor is the widespread use of artificial insemination by which 60–70 per cent of our dairy cattle are now bred, and in Denmark the figure approaches 100 per cent. Here a group of decision makers have emerged, with power to implement these decisions, and this group is different in both training and background from the traditional breeders. Widespread milk recording in this country provides the necessary information. Many geneticists then see future breeding schemes for dairy cattle as centred round the extensive use of bulls which have been evaluated from the performance of their daughters in commercial milk producing herds. Because they may have hundreds of daughters by artificial insemination, such bulls can be much more accurately judged than has ever been possible before. Interestingly, some individual breeders are now forming groups to compete with the artificial insemination organisations on their own ground, a very encouraging development.

In pigs the classic application of rational methods of improvement is in Denmark where co-operative stations for testing boars on the growth and carcasses of their progeny have been an integral part of the breeding programme for sixty years. In the last five years in this country we have taken a further step forward. Realising that many characteristics that we want in pigs are measurable on the boars themselves, 'performance and progeny testing stations' have been set up at which about five thousand boars are tested each year for their own growth rate and fat thickness and for carcass measurements on some of their brothers and sisters. These new developments in pig breeding put us, probably, years ahead of any other country.

In poultry breeding, more especially for meat, we have the clearest evidence of the effects of modern breeding programmes as seen in our own homes. Compare in your mind the turkey that you ate last Christmas with those you were eating only ten years ago. Chickens too are no longer a special treat but an everyday food and a lot of the credit for this must go to the breeders and to their professional geneticists.

How are these animals bred? Any breeding programme which leads to intense concentration on the blood of a few genetically superior individuals causes inbreeding and this almost invariably leads to a decline in vigour, most obvious in the ability to survive difficult conditions and to breed. Although it does not matter how many eggs the turkey you ate might have laid, it still has to hatch from an egg and so the vigour of its parents is important. If we cross two different highly selected lines, both of which have been closely bred, the cross-bred progeny will recover from the inbreeding degeneration but retain the effects of selection. Cross-breeding of some kind is then important in breeding most of our animals, with the exception of dairy cows. Your turkey, for instance, will probably have a father who has been bred for growth rate and breast muscles and very little else – most commercial turkey breeding is done by artificial insemination in any case. He will have been mated to a hen who will be a cross between two different strains, selected not so intensively for growth rate and muscling but also for egg production. She will have some hybrid vigour for egg production – this means that the turkey chick will be fairly cheap – and she will also pass on to it some genes for growth, which will be greatly added to by the genes that it gets from its father. Whether or not we use crossing depends on how expensive it is to keep as sources of useful genes these strains which are in themselves inferior. Even the poorest strains of turkeys lay so many eggs that this is not too expensive, but it is quite out of the question in cattle where, on average, a cow produces only two female calves in her whole life. As I mentioned earlier, future selection programmes in dairy cattle will rely mostly on our ability to identify superior bulls in artificial insemination. We should be able to find enough such bulls every year to avoid too much concentration of blood. In fact, the rate of concentration in artificial insemination in Denmark is proving to be less than in days of natural service.

What of the future? I take the view that our knowledge

of what to do is in most cases far ahead of our opportunities to do it and so I should be very happy merely to see many of the ideas, which have come to the fore in the last ten years, put into practice. I would hope that we would analyse such breeding programmes as they went along and find out quickly whether or not some of the ideas were working. But in many ways we are still at a fairly primitive level. I spoke earlier of the process which started two hundred years ago of sorting out differences between the existing local varieties of the animals. Very surprisingly this has still not been done adequately, even in this country. We have imported breeds from abroad that seemed likely to be useful and indeed have been, whereas we are still not sure whether some of our own breeds, particularly of beef cattle, might not do equally well and there are still other breeds abroad which we might usefully try out.

The spectacular expansion of molecular biology in the last ten years has had little or no impact on animal production and I don't see any immediate likelihood of this. But it must very greatly alter our general approach to our problems, and to suggest kinds of questions which we wouldn't have formulated in the past. I would particularly like to know exactly what, in fundamental biological and genetic terms, makes the Friesian cow which when fed properly produces a lot of milk but with a moderate fat content, differ from a Jersey which produces a moderate amount of milk with plenty of fat, or from an Aberdeen Angus which produces almost no milk but muscle with plenty of fat within it, or from a Charolais which produces a lot of lean muscle. These may not seem to be genetical questions, but I suspect that only with the help of geneticists will they be answered. We have almost no clues to the answers at the moment, and the questions clearly are also relevant to human problems, such as why some people get fat and others not.

The most interesting technical developments will come from experiments on fertilised eggs, a line of work being actively pursued in Cambridge. It is now simple to transplant fertilised eggs in both sheep and pigs, but it is much more difficult in cattle. This is the parallel to artificial insemination on the female side. Up to twenty fertile eggs, produced by mating two superior animals, could be got at the same time and transplanted into perhaps twenty different recipient cows. We could then test the breeding ability of cows by the way in which their daughters perform as we do with bulls at the moment. But there are more interesting possibilities. One which is certainly not very far away is to determine the sex of the fertilised egg while it is in transit between donor and recipient and depending on what one wants, produce offspring of one sex only, but Dr Edwards will discuss this in more detail in the next article. Although this is perhaps limited in its practical application it seems much more possible than trying to divide semen samples into male and female types by, shall we say, putting them in an electric field to separate groups all of which bear X chromosomes, which will produce females only, or Y chromosomes which will produce only males. The scientific literature as well as the Sunday papers have not been short of such claims, but not one has stood up to repeated experiments. There is also the possibility of taking a fertilised egg at, say, the eight cell stage, separating the eight cells and having them developed as eight separate but identical individuals, a set of identical octuplets. This still remains merely a good idea. One might also make genetically identical individuals by transferring not single cells, but individual nuclei (which contain the genetic information) from one individual into the cell of another, possible so far with newts but not with anything higher.

Let me put aside speculation and deal with the relevance of all this to the problem facing the world today – whether we can so control the breeding and feeding of the human species that there is any room for animals on earth at all. In the developing countries, the problems in animal breeding are not sophisticated – they are mainly to decide which existing breeds and crosses between them are best suited to the local environments and exactly how they should be used. Such decisions can often have major effects on the efficiency of animal production. In countries with a more developed agriculture, I think we can also look forward to steady, if perhaps not spectacular, progress.

3 Reproduction: chance and choice

R. G. Edwards

The control of parents over the kind of offspring they beget is almost non-existent. This situation is accepted without question by most people, and any thought of controlling the quality of their offspring is completely novel. It is only when the occasional deformed or genetically abnormal baby arrives that most people realise – and almost always painfully – what a lucky or unlucky dip procreation really is. The birth of a deformed baby is among the most distressing situations in medicine.

We have come to recognise, especially in the last half-century, the importance to the individual of the period of life before birth and to realise how little we can do about exerting any control over prenatal existence. We are all familiar with the effects of German measles on foetal development, and most people accept that an abortion is now preferable to the birth of a deformed child. Yet the moment that this principle is extended to other foetal deformities, or to techniques other than abortion, doubts and uncertainties arise.

In relation to reproduction many people have the feeling that foetuses 'should not be tampered with'. There are obviously more serious grounds for concern than just this.

The stages of life shortly before and after fertilisation encompass events that have profound effects on the future child, and a great deal of study has been devoted to these stages of development in animals. Curiously enough, however, very few of these studies have been devoted to studying or curing human deficiencies that arise early in development. So, when attempts are made to report actual or potential progress for medicine in this area, it is all too easy to equate knowledge obtained for scientific reasons with dreadful consequences that would follow its application to man. Few commentators seem to bother with the beneficial results that could arise from this knowledge. Quite obviously, before evaluating progress in this area of research, clearly we must be aware of both the scientific problems and opportunities, and also of the objectives that are immediate and urgent for medicine.

First consider the scientific side before returning to the medical aspects and to the social, ethical and other issues which are raised by these experiments. A major scientific problem is how to grow eggs and embryos under controlled conditions, and this means that we must learn how to culture them in the laboratory. We know how to obtain and handle mammalian eggs, and some aspects of their metabolic requirements have been analysed. Eggs in their earliest stages of development can metabolise only a limited number of nutrient chemicals which must be provided in the culture media, and after their culture through parts of their early development many will develop to full term following transfer into a recipient female. But there is a lot to learn about the culture of these eggs for our techniques are not very successful, even with the earliest stages of embryonic development. Another necessity is to achieve routine fertilisation of the eggs in the laboratory. Spermatozoa taken from a male cannot fertilise eggs immediately, because they have to go through a 'ripening' process in the female tract. At the moment, it is possible to fertilise animal eggs of some species by taking spermatozoa from the female tract, or using secretions of the tract to ripen the spermatozoa. It is difficult to achieve fertilisation in animals, where the system can be brought under some degree of control, although discoveries are now coming rapidly. In man, too, fertilisation should soon be a routine procedure.

Despite the difficulties of culturing embryos, a great deal of scientific knowledge has been gained. One conclusion of major importance that has emerged from all these studies is that in the early stages of the development the embryos have not yet become fully organised. They thus tolerate manipulation such as the injection of substances, the transfer of nuclei from other cells, the fusion of two or more embryos together, the splitting of one embryo into two or more parts, and the inclusion of other cells into the embryo. Obviously, the application of these methods has considerable effects on embryonic development, ranging from the production of twins or chimeras to 'vegetative' reproduction whereby many identical embryos can be produced by nuclear transfers from a single donor. Experiments of this type give us invaluable scientific knowledge of early development.

We also know, from biochemical and embryological studies that the genes carried by an embryo are not working immediately after fertilisation, for there is a delay until the embryo has approximately sixteen cells. Some mouse embryos carrying a mutant gene die at this stage of development. Many further questions have arisen as a result of these and other studies – for example, what are the properties of individual embryonic cells, or how can cells just beginning their differentiation be obtained for analysis.

Having dealt briefly with the scientific side, I shall now turn to the totally different area of the medical aspects of this work. There are three major areas of medicine which could benefit enormously from these studies. First, some forms of human infertility could possibly be cured; second, knowledge gained could be useful for the development of contraceptives; and third, knowledge and methods could be obtained leading to the alleviation of genetic or other deformities. Each of these areas demands that we know in detail the events during conception and development of the embryos until they implant in the womb.

When we can take human eggs from the ovary, prepare them for fertilisation, ripen spermatozoa, and then fertilise and grow the eggs through their early stages, we shall be in a position to attempt to cure some forms of infertility in women. A major cause of infertility is a block in the tube called the oviduct – this prevents the ovulated egg from entering the uterus. Once blocked, the oviduct rarely can be repaired. The cure of this sort of infertility would, therefore, demand that hormones be given to the woman to stimulate the eggs, followed by a minor operation to obtain them from her ovaries. Spermatozoa would then be taken from the husband. No major problems arise so far, but the development of eggs outside the mother, and their transfer back into the mother again will create difficulties, at least for some time. Despite our knowledge of fertilisation and cleavage in animals, the best place to achieve these events is still in the oviduct. At one time, the co-operation of another woman was thought to be necessary. For example, a woman due for an operation for hysterectomy could nourish the embryos in her oviduct through these difficult stages for a day or two, before her own oviduct is removed and the embryos recovered. Foreseeable advances in technique should avert this necessity.

Medically, there would be few problems to overcome. Ethically, a number of points must be raised. One of these is that we would have to take several eggs from the mother, and transfer only one or two back into her. The remainder would be thrown away. Is it acceptable to discard the excess embryos? Let us remember that these embryos are two to three days old, the size of a pin-head, have not begun to differentiate and are composed of less than a hundred cells. If anything can be certain in this field, it is that no organs, nervous tissue, or sensory cells have developed, or will develop for many days. Another point is worth remembering in this connection. We now suspect that women who use an intra-uterine device are protected from pregnancy because this device causes the death or expulsion of an embryo at exactly the same stage of development that we would transfer them in our work. So, in principle, by throwing the excess embryos away, we are doing no more than any couple using the intra-uterine device for contraception. I believe that we are fully justified in doing this, even if only to help couples who cannot have children of their own. In a society which sanctions the abortion of a fully-formed foetus, the discarding of such a minute, undifferentiated embryo should be acceptable to most people.

Apart from infertility, we would learn a great deal that

could help with our next major aim – knowledge for the development of contraceptives. We know so little of reproductive events in humans – and indeed, our knowledge of some events in animals is meagre. For example, how do spermatozoa ripen in the female tract ? This is a fundamental necessity for fertilisation to occur, and could be an excellent and acceptable target for contraceptive measures. We also know next to nothing in man about the movement of the embryo down the tubes, yet in animals this movement is profoundly accelerated by the levels of steroid hormones and leads to the expulsion of the embryos. This approach, indeed, could and is being utilised in the development of a pill used for contraceptive purposes after intercourse has taken place. We know very little in animals, and nothing in man, as to how the embryo implants in the uterus. Can we, for example, immunise people so that fertilisation or implantation are prevented ?

I shall now turn to our third major problem – the alleviation of inherited human deformities. Some of these are extremely distressing, and at the moment, there is little that can be done about them. We need much more information than we now have on the origins of some of these abnormalities. One such disorder is Mongolism, of which there are three major types. The most common type has an extra chromosome in all cells of the body, and we now consider that the origin of this type of Mongolism is due to an abnormal event before or after fertilisation. It is most important to decide on the origins of this disorder, for this form of Mongolism occurs sporadically and unpredictably, although it is most common in children of older mothers. We will obviously be in a better position to prevent this type of Mongolism when we know the cause.

The second form of Mongolism ('translocation' Mongolism) though rare, is strongly inherited. Here, the extra chromosome or most of it, has become attached to another chromosome. This presents us with a different problem; we have to type the embryos, identify the abnormal ones and discard them. Provided embryos can be typed in this way, this method could be applied to any genetic character identifiable in the early embryo. The question boils down as to how the necessary methods for identifying characters can be devised. One approach is to remove a few cells from the early embryo, and use them for typing; rabbit, mouse and sheep embryos can tolerate this form of interference and continue their development normally. These techniques will not be easy on man, but they are by no means impossible. Apart from translocation Mongolism, we might ultimately detect other chromosomal errors, and also the sex of the embryo. Sexing could offer a means of averting the birth of children with such diseases as haemophilia, a form of muscular dystrophy, favism and other defects. In the inheritance of these diseases, a mother who is a carrier of certain sex-linked genes, the foetuses are examined at three months of gestation and, if the mother agrees, those carrying the gene are aborted. The risks of interfering with the pregnancy at this stage, and the need for abortion, make this an unsatisfactory approach. It would be far more acceptable to tackle these problems at their source – i.e. during early development.

The second great advantage of this kind of treatment is that where the carrier of the genes can be identified the family would be cleared of these afflictions for several generations, until either a normal gene is mutated – a very rare phenomenon indeed – or descendants of the family married into another family with a similar affliction. It must be seldom that an opportunity is presented whereby advantages like these can be given to generations to come. There is one major ethical problem here – it concerns controlling the sex of children in order to avert inherited diseases. There is a built-in safeguard against using this method for social reasons – the necessary techniques are not simple, so their widespread use would probably be precluded. Yet, once these methods were perfected, there would be a temptation to apply them to gratify the needs and wishes of parents.

These, then, are the medical reasons that lie behind our

work. They are a long way from the daunting prophecies made in some quarters. Among these sensational prophecies are the production of identical 'clones' of children by means of nuclear transfer, the 'building-in' of genetic information into embryos and the coding of future behaviour. Another hardy perennial is the growth of 'test-tube babies'; if this means growing the embryos to a recognisable human foetus in a test-tube, then there will be a long time to wait, for we are still at the very beginning of studies of this kind in the most co-operative animal species, and we know merely enough to realise how difficult this will be. With time, many of these prophecies will undoubtedly be realisable and some could be beneficial to society, but we shall have had ample time to decide in which way such discoveries should be used.

I believe that our work can prove beneficial for a large number of people. A contention could be raised against curing infertility in that we would further increase the pressure of over-population at a time when most efforts are aimed at contraception. This argument must be given short shrift – we cannot selectively penalise the infertile for this reason. Infertility can give rise to various problems, and adoption is often no substitute for one's own child. These are hard facts to live with and hesitation and delay in alleviating them increases the burden. Another charge is that the methods we propose to eliminate genetic defects will be too difficult in practice and help too few people. But advances come quickly, and what is difficult today is easy tomorrow. Another consideration is that some couples are infertile for genetic reasons, and by curing their infertility we would increase the incidence of the deleterious genes. We must obviously give attention to this point.

For us, there are sufficient major decisions to be faced for some time to come. A major example will be the decision to transfer embryos into the mother for continued development for it will be difficult to ensure that the embryos are developing normally. We recognise that work in this field is of interest to a great many people, and not only those directly affected by it. Recent developments in other areas of medicine have shown this to be true in transplantation surgery. While scientists and doctors will obviously have to lead the way and point to the possibilities of raising the quality of human life, they should not be asked or expected to answer alone the other critical and diverse questions that arise from their work.

Equally, irrational pessimism and doubt will interfere with true judgement. The only correct approach lies in a critical evaluation of the scientific and medical problems that exist, the benefits that could accrue from their solution, and the realisation of the involvement of many people in the moral responsibility that accompanies work on human conception.

4 The chromosome construction and reconstruction of wheat

Ralph Riley

More than ten thousand years ago, in the Eastern Mediterranean, genetic changes occurred to certain wild grasses that were of profound significance for the subsequent history of man. The wild grasses concerned were the ancestors of our cultivated wheats and the effects of these changes are still present in the wheats of today, where they both help and hinder research workers who are trying to breed better varieties. Why I said that these changes played such a significant part in our history is that without them wheat could never have become a crop, since it would not have been capable of exploitation by the hunters and gatherers of Western Asia when they first adopted a settled, agricultural way of life. Clearly the origins of agriculture depended on the availability of food plants that were suited to domestication from the outset and that were capable of continuing adjustment to the needs of the growers. There were relatively few appropriate species, but wheat fitted the role perfectly. It sustained the agricultural revolution of the neolithic period and was ultimately largely responsible for the provision of the food surpluses that gave man time to contemplate, to invent and to build towns, tombs and armies.

The genetic changes affected the chromosomes of wheat and the way that these chromosomes behave in cell division. But before I come to the special significance of chromosome behaviour for the history of wheat and for its future, I shall talk briefly about the way that genetics is used in plant breeding.

Plant breeding consists essentially of the development of forms of cultivated plants that are better adjusted genetically to the needs of the grower, user or consumer, than those previously available. In breeding work, the genetic structures of plants are modified so that, when expressed physiologically or morphologically, they lead to higher yields, the reduction of losses from such hazards as diseases and pests and improvements in the quality of the harvested produce.

Rule-of-thumb methods of crop improvement were used long before the advent of the genetic era. The oldest method, which is still employed, relies on selecting from mixed populations those plants best adapted to agricultural exploitation. Unconscious selection was responsible for the very early evolution of our crop plants, while conscious selection has been responsible for their detailed adjustment and adaptation to our own more specific needs.

But, towards the end of the nineteenth century, the efficiency of selection had improved greatly and crop populations became much more uniform. Now obviously selection is not possible in a uniform population, so it became necessary to generate new variation. Several nineteenth century plant breeders recognised that new variation, on which selection could be practised, occurred in the progenies of hybrids between different agricultural forms or varieties. As a result, hybridisation and selection were used in a complementary manner for some time, even before a scientific explanation could be offered for their successes.

Following the rediscovery of Mendel's Laws in 1900 and the rapid development of genetics in the early years of this century, the results of hybridisation and selection became explicable in scientific terms.

Later on, with the recognition of genes controlling useful characters, it became possible for the breeder to produce hybrids that gave him a good chance of isolating new combinations of beneficial genes. In this way, genes giving resistance to different diseases could be brought together and at the same time combined with others controlling, for example, such characters as height or earliness. In addition, these genes could be set in genetic backgrounds that favoured the production of high yields of produce of good quality.

One of the major steps in our understanding of heredity came with the realisation that genes are linked together on chromosomes and that, in any species, there are the same number of linked groups of genes as there are pairs of chromosomes. This is part of the evidence that genes do, in fact, reside on chromosomes.

The recognition of gene linkage, and of the significance of chromosomes, stressed the importance of the process of meiosis – the cell division that precedes the formation of eggs and pollen. Linked genes generally become arranged in new ways by the genetic crossing-over, or recombination, that takes place during meiosis in hybrid individuals. But I shall explain this in more detail.

In the cells of higher organisms there are varying numbers of pairs of chromosomes – twenty-three in man, twenty-one in bread wheat and so on. The number of pairs is constant for all the cells of an individual, and more or less constant for all the members of a species. One member of every pair is derived from the male and one from the female reproductive cell from which every individual arises on fertilisation. Because they may come from separate parents, partner members of a pair of chromosomes may have, at several places along their lengths, alternative versions of genes giving different expressions of characters.

During the first part of the meiotic division every chromosome becomes 'paired', with its partner. At this stage breakage and joining takes place between the paired chromosomes in such a way that segments are exchanged. These exchanges can ultimately be recognised, in the progenies of hybrids, to have given rise to changes in combinations of previously linked versions of genes. After this the paired chromosomes separate, passing to one or other of the two cells that are produced in the first part of the division.

As I have said already, the plant breeder aims to rearrange the arrays of genes in the hybrids he makes. If the genes are very tightly linked together, this can seriously impede the breeder's attempts to produce new arrangements.

This limitation on the rearrangement of genes has posed a problem for a long time and plant geneticists have wondered how they could increase the rate of release of genetic variation from hybrids, so that new patterns of linked genes could be produced. There are two ways in which such changes might be encouraged. The first is by increasing the frequency of exchanges between already paired chromosomes and the second is by improving the capacity of chromosomes to pair.

In fact, my own work has been concerned with the second possibility – that is, changing the chromosomes capacity to pair.

At the Plant Breeding Institute, we have been able intentionally to interfere with normal chromosome behaviour to produce new arrays of genes in wheat. This involved changing the mode of chromosome pairing in the first part of meiosis and we used this technique to introduce into bread wheat an unusual form of disease resistance.

To explain the work, I shall go back to the genetical changes in the ancestors of wheat that I mentioned earlier. The 42 chromosomes normally present in each cell of wheat, were derived in evolution from three wild grasses of the Eastern Mediterranean. These three species each have 14 chromosomes and the three sets of 14 were assembled together in bread wheat in the following way. First a form of the most primitive wheat species which is nowadays widespread in the Middle East, hybridised with a species of goat grass that today extends from the southern shores of the Black Sea across central Turkey into Syria and Israel. The resulting hybrid also had 14 chromosomes, but by an accident of cell division, the chromosome number doubled to give rise to a new 28-chromosome wheat called 'emmer'.

If the chromosome number had not doubled, this hybrid would have been sterile because of the irregularity of chromosome behaviour at meiosis. But in the 28-chromosome wheats, every chromosome has a partner, so chromosome behaviour is regular and the eggs and pollen are fertile because of their balanced constitutions. Twenty-eight chromosome emmer wheats were also wild growing species, but they were taken into cultivation, together with the 14-chromosome forms, by the earliest farmers. Remains of both of these types have been found by archaeologists at the earliest levels on sites, like that at Jericho, with the longest histories of habitation. Their much changed derivatives are still cultivated today – for example, the wheats used to make macaroni and spaghetti have this chromosome constitution.

Subsequently, another goat grass that at present grows in a broad band stretching from the eastern end of the Black Sea into Kashmir and the north of West Pakistan, hybridised with the 28-chromosome emmer. Again the initial hybrid, which had 21-chromosomes, would have been sterile but hybridisation was accompanied by chromosome doubling. This resulted in a new 42-chromosome species and from this our present bread wheats arose. At this chromosome level, wheat had great evolutionary flexibility and as a result it has become the principal food crop of vast areas of the world.

In the genetic structure of 42-chromosome wheat, every chromosome has a full partner with which it pairs at meiosis and this partner was derived in evolution from the same wild grass parent. If you like, there are the chromosomes of three distinct species of grass in each wheat cell. Because of the three ancestral sets of chromosomes many genetic activities are triplicated. This is a feature that has made wheat particularly suited to experimental manipulation by geneticists and also so amenable to cultivation as a crop plant.

You would think that there would be a confusion of pairing at meiosis because chromosomes derived from one ancestral species would pair with the corresponding chromosome from another. But this does not happen, and the reason is due to genetic activity of one pair of the 42-chromosomes. This pair is known as 5B. It was because of the origin of this activity on chromosome 5B that wheat became suited to agriculture ever since neolithic times.

We have used knowledge of this activity in improving wheat and I shall now describe how we were able to exploit it to transfer resistance to a fungal disease called yellow rust into wheat from a wild grass.

The most economic means of limiting losses due to diseases of crops is to grow forms that have genetically determined resistance to infection – that is to say on which either the disease does not develop or does not do so in a damaging way.

Most varieties of wheat that are grown in areas in which yellow rust is a danger, have some measure of protection because of genetic resistance. If they did not have this in-built protection, the crop might be destroyed. But it is necessary continually to seek new sources of resistance because of the risk that genetic changes in the fungus may cause it to become virulent on forms of the plant that have previously been resistant. Indeed, in the past two or three years there have been dramatic changes in the yellow rust fungus in Britain and the resistance of several varieties of wheat has been overcome.

There are wild species of grasses related to wheat that have desirable levels of resistance to yellow rust and we wanted to incorporate this resistance into our cultivated wheat. We were particularly interested in the grass *Aegilops comosa* which grows in the Aegean area. The simple approach to the problem would have been to cross *Aegilops comosa* with wheat – but when you do this, you find that its chromosomes do not pair up with wheat chromosomes at meiosis – and so gene exchange cannot take place and the resistance cannot be transferred into the wheat in a straightforward way.

But we realised that if corresponding chromosomes from differing ancestral species of wheat were prevented from pairing by the activity of the chromosome 5B, then maybe the same action prevents pairing between *Aegilops comosa* and wheat chromosomes. We decided to test this idea and the test was carried out first of all by a backcrossing programme. In backcrossing, a hybrid and its derivatives are crossed repeatedly to one of the parents, known as the recurrent parent. In this way, we can introduce into the recurrent parent simply inherited single characters. We used the backcrossing programme with our *Aegilops* wheat hybrids and we were able to produce plants that had all the normal wheat chromosomes and an extra pair of chromosomes derived from the *Aegilops* species. These extra chromosomes carried the rust resistance genes. During the backcrossing programme we inoculated the wheat seedlings in each generation with a large dose of yellow rust and after ten to fourteen days, we selected those that were resistant to attack.

So, by backcrossing and selecting for rust resistance, we were able to isolate the line in which the pair of immigrant *Aegilops* chromosome was present. But this was not good enough since plants with this additional pair are abnormal in various ways. So, we were faced with the problem of interfering with the genetic restrictions of the chromosome pairing so as to give the immigrant chromosome a chance to pair with a related wheat chromosome. This we did by using another wild grass which we knew had the capacity to suppress genetically the regular meiotic chromosome behaviour of wheat. The idea was that if we could stop the 5B activity it would enable the *Aegilops* chromosome segment causing rust resistance to become incorporated into the wheat chromosome set. The wheat line with the additional chromosome was hybridised with this goat grass and again we backcrossed with wheat as the recurrent parent. And once again, the seedlings were infected with rust and we selected those that were resistant.

Two years later, after four generations, we were able to isolate plants with 42 chromosomes that showed the same degree of yellow rust resistance as our original Aegean grass. Twenty pairs of chromosomes of this line correspond precisely with those of standard forms of wheat, while the twenty-first pair has a wheat segment and an *Aegilops* segment. The *Aegilops* segment carries the gene giving rust resistance. The chromosome on which it is carried was the product of an exchange that occurred while the 5B activity was suspended. Apart from its resistance – and it may be said that it is resistant to every form of pathogen to which it has been exposed so far – the resistant line is no different from the recurrent wheat parent. Breeders in several countries are now transferring this resistance to commercial varieties for agricultural exploitation. The results of our work with the rust resistance of *Aegilops comosa* demonstrate the validity of the ideas on which it was based and show that new ranges of genetic variation can be exploited in wheat improvement. The work emphasises the benefits that may follow from intervention in the course of meiosis.

Ultimately the kinds of genetic variation affecting meiosis that I have discussed in wheat will very likely be used to provide an understanding of the causal basis of the process in biochemical terms. When such an understanding has been attained we shall more readily be able to intervene experimentally in the sequence of meiosis. This will assist plant and animal breeders and may indeed throw light on other biological problems like, for example, some of the causes of human reproductive errors.

As an applied biologist, the peculiar attraction that I find in looking at our present work on wheat is the lengths of the vistas that we can survey. Looking backwards I can resolve more sharply the events that influenced the first applied biologists – the neolithic farmers. Looking forwards, I can discern the ways in which my successors will introduce even greater precision than is yet within our grasp.

5 Genetic counselling

A Physician

Other talks in this series have dealt with some of the interesting applications of genetics that we may expect to see in the future. The subject I want to discuss, however, is an application of genetic knowledge here and now. This is the provision of genetic advice for those who need it. It is, in my opinion, much the most important single practical application in the human field, as things are at present.

The proportion of people in the population who really need genetic advice is fairly small. Hazarding a guess, perhaps one person in twenty or thirty might profit from genetic advice at some time or other during their lives. But though the proportion is not large, many of those who need advice, need it very badly. For example, the birth of an abnormal child is a terrible shock, and parents will often want to know what the risk of recurrence may be should they have another child. Unfortunately, the whole subject bristles with old wives' tales, and old wives' tales do not often err on the side of foolish optimism. The unfortunate couple, if they do not seek doctors' advice, will often get plenty from other and very ill-informed sources. But, the steady growth of human genetics over the years has provided a wealth of data which makes it possible to give useful advice in a very high proportion of cases.

Just what are the reasons for seeking genetic advice? A recent analysis of cases seen over a period of fifteen years at one clinic showed that 85 per cent of all inquiries came from couples who had had a child suffering from some defect or other and who wanted to know what the risk of recurrence might be for subsequent children. The balance was made up in almost equal numbers of those who themselves suffered from some defect which they feared they might pass on, and those who had something in their family history which might imply special risks for children. This is likely to be the pattern of inquiries for some time to come.

Giving genetic advice is much facilitated by a very special advantage. There is a two-way effect. On the one hand we have defects which are simply inherited, or fairly simply inherited. Then, with little that is intermediate, we pass to conditions which are certainly not simply inherited, but where it is clear that inherited constitution plays a part in causation. This is shown by an increased incidence of the particular condition in close relatives as compared with its incidence in the population generally. As a very general rule, we can say that the simply inherited conditions carry a high risk of recurrence, like 1 in 2 or 1 in 4, and so on. So, when the outlook is gloomy the adviser usually knows just where he is. But with the partly inherited conditions, although the exact genetics may be rather obscure, the chances for the inquirers are usually fairly good, indeed, sometimes very good. Here we are dealing with risks of recurrence of, say, 1 in 20, or 1 in 30, or even less. In fact, if we say that a risk of recurrence which is worse than 1 in 10 is a bad risk, and a risk of recurrence which is better than 1 in 20 is a good risk, then in practice very few estimates fall in between. In a word, the risk for further children tends strongly to be either bad or good.

Now, let us take the simply inherited defects first. They are almost always individually rare or very rare. It is true that there are some simply inherited conditions, which because of certain associated advantages, have become common in some parts of the world. A notable example is sickling and sickle cell anaemia in most of Africa. But there are not many of these conditions and I will not mention them further, because they will be dealt with by Dr Allison in the next article. There is no such common harmful gene in our own population. With us, the commonest simply inherited defect is cystic fibrosis, a defect of the mucus producing cells, which leads to abnormalities in the digestive and respiratory tracts. It affects one child in every 2,000. This, however, is common enough to have made it worthwhile setting up the Cystic Fibrosis Trust, which does excellent work in keeping parents informed and in encouraging research.

Most of the other simply inherited conditions are much rarer than this, and some, in fact, are very rare indeed. But I must not under-rate their importance, because although they are individually rare, they are important in the aggregate. This is because there are so many of them. In a splendid encyclopaedia, published in 1967, Professor Victor McKusick, of Baltimore, lists no fewer than some 700 defects whose mode of inheritance is considered proved, together with another 800 or so where there is very suggestive, even if not quite conclusive, evidence.

The chromosomes are paired, so the genes are paired. There are two of everything. Some defects are dominant: that is, the abnormal gene is dominant and the normal alternative gene, is recessive. So a person who has one abnormal dominant gene is abnormal. This leads to the picture, as we see it in families, of direct transmission from parent to child, as, for example, with Huntington's chorea, a slow degeneration of the nervous system. Every affected person has an affected parent, because the abnormal gene must have been received from one parent or the other, and that parent, of course, must have been affected. An affected person has one abnormal gene, and one normal gene, so when he or she marries a normal person, on the average, half the offspring are affected. Some of these dominant defects have been traced through ten and more generations. But there are complications. One is that we may get back to the original mutation when the abnormal gene first appeared. If we do, persons of previous generations are all normal. A notable example is the commonest type of dwarfism, achondroplasia, which is familiar to everyone. Actually, the great majority of such dwarfs are born to normal parents; it is a newly occurring mutation. But, of course, when a dwarf marries a normal person the chance is 1 in 2 that any child will be affected in the same way. With dominant genes in general the chance that an affected person will have an affected child is 1 in 2. But there are other complications; for example, there is often great variability. Some people may be severely affected,

43

others only slightly. This, and a good many other factors, must be taken into account in giving genetic advice.

Then we have all the recessive defects; I have already mentioned cystic fibrosis. Here the abnormal gene is recessive and the normal alternative dominant. People who carry one abnormal gene only are outwardly normal, but, should two such carriers happen to marry, there is one chance in four that any particular child will get the abnormal gene from both parents and so be affected. Usually we find an entirely negative family history, but often more than one child in a sibship of brothers and sisters will be affected. The chance, as I have just mentioned, is in fact 1 in 4. So, suppose we are consulted by the parents of a child suffering from cystic fibrosis. We have to tell them that the chance that any further child might be affected in the same way is 1 in 4. But we can add quite a lot more. We can stress that all of us, without exception, carry a few harmful recessive genes; the usual estimate is about five per person, or their equivalents. This is because harmful recessive genes are not very uncommon. What is uncommon is their coming together in pairs, and so producing an outward effect. Most of us are lucky; we happen to marry someone who does not carry our own particular brands of harmful gene, but quite different ones instead, so they can never come together in pairs. The couple with an affected child are no different from anyone else. They have just been unlucky because they both happened to carry the very same one. So, I hope, we can help to dispel feelings of guilt, or of being different from other people. Then we can reassure them about the normal children they may be lucky enough to have. Such normal children have nothing to fear when the time comes for marriage and children. True, such normal children have two chances in three of being carriers, but, provided they do not marry a cousin, the risk of marrying another carrier is so remote that they can forget all about it. We can also add a similar word of reassurance about other relatives, such as brothers and sisters and nephews and nieces of the couple.

44

Then there are the sex linked genes, those carried on the X-chromosome. These are especially important in men. McKusick lists about 80, including such important conditions as haemophilia and the commonest kind of muscular dystrophy. With these two conditions the gene is recessive. A carrier woman is outwardly normal, but on the average, half her sons, who have only one X-chromosome, will be affected, and similarly, on the average, half her daughters will be carriers like herself. This is the clear picture we see of something carried by women and manifested in men.

An important field, in which advances are continually being made, is the detection of carriers of recessive genes. It is being found in one condition after another that although the carriers display no apparent abnormality, sophisticated biochemical tests may show deviations from the normal. The detection of carriers is especially important with sex linked genes. So, with the commonest form of muscular dystrophy the prior chance that a sister of affected boys carries the gene is 1 in 2. A few years ago, this would have been all we could say to such women, namely, that the chance of any boy being affected was 1 in 4. But, of course, if the truth could be known the real chance for half the sisters is 1 in 2 and for the other half effectively nil. Affected boys have very high values of a breakdown product of muscle, creatine-phospho-kinase, in their blood. Then it was discovered that the great majority of carrier women have creatine-phosphokinase values above the normal. So it is now possible to test the sisters of affected boys and in 90 per cent of instances they can be told either that unfortunately they do carry the gene, or else that they have escaped getting it, and so their children will not be at risk.

At this point we might pause to ask why the simply inherited defects are always, unless there is a counter-balancing advantage, rare or very rare. It is because mutation is rare, and natural selection sees to it that the frequency of such defects in the population is kept to a low level. But this is another topic, to be dealt with more fully by Dr Allison.

When we turn to the partly inherited conditions, the picture, as I mentioned earlier, is quite different. Pride of place, in terms of numbers, goes to the commoner congenital malformations, for example, the major malformations of the central nervous system: a failure of development of the brain called anencephaly; spina bifida, a failure of the neural tube to close properly; and hydrocephalus. Then there are such conditions as the great mass of severe mental retardation of no specific type that we can recognise. There is congenital heart disease, harelip and cleft palate, clubfoot and a good many more. These are all commoner than the simply inherited abnormalities. They just cannot be simply inherited; natural selection would see to that. But there is a constitutional genetic element. Thanks to many large-scale surveys we know the approximate risks of recurrence if the couple should have another child. With anencephaly or spina bifida it is about 1 in 25, and the same is true of harelip, with or without cleft palate. With congenital heart disease, lumping all forms together, it is about 1 in 50.

Now, at a genetic clinic I always take the view that it is not our job to tell potential parents whether or not they should have children. It is our job to explain the risks as clearly as we can, and then the decision must be left to the couple themselves. We find that many, perhaps most, couples, when given a bad risk, such as 1 in 4, do in fact decide to avoid further parenthood. Many will adopt a child instead, and this seems a wise decision. But with the low, or fairly low, risks we can offer yardsticks. The chance that any random pregnancy will end with a child suffering from some severe malformation or other, or that some serious error of development will manifest itself in early life, is probably little less than 1 in 30. If the special risk for the particular couple is not out of proportion, it does not seem to be a risk of such an order that sensible people, knowing the facts, need be deterred. After all, if parents were to be deterred from having children by very small risks, no one would ever have a child. We do in fact find that very many couples, having thought things over, decide to take the small risk and go ahead.

I must not give the impression that genetic counselling does not have its complexities. For example, what appears to be just the same end condition may be due on different occasions to quite different genes, and indeed sometimes to non-genetic, that is, environmental, causes. If we take deaf-mutism, the majority of instances are genetic, and when this is so, are usually due to recessive genes, though by no means always to the same recessive gene. There are many recessive genes, any one of which, when present in double dose, determines deaf-mutism. Then, a very few instances are due to dominant genes, and a very few to sex linked genes. In a fair proportion of instances deaf-mutism is non-genetic, though usually we cannot identify the environmental cause. But knowledge is growing, and all the time researches and surveys are being carried on which lead to clarification and simplification. It is possible, in the present state of knowledge, to offer useful advice in a high proportion of instances, and this precision will be continually improved.

One increase in knowledge, which is continually being made, is the separation, into distinct and recognisable groups, of conditions which before had to be lumped together. This, in fact, has been the pattern of progress through the years. Let us take severe mental retardation as an example. The great bulk of cases fall into the broad undifferentiated category. Rather more than thirty years ago phenylketonuria was separated from the rest. This is a metabolic defect due to a recessive gene. It is, as usual, a rare condition, but it is well worthwhile screening new-born babies because early recognition, and so early treatment, greatly improves the chances for more or less normal development. Over large areas of the country this is being done as a routine test. One after another, such recognisable entities are being separated off. Less than ten years ago workers at Belfast and in America spotted a particular kind of mental deficiency and found it was due to another metabolic error; it is called homocystinuria. And, typically, once it had been spotted, those at hospitals for mental defectives realised that they had been seeing these cases for years.

There is undoubtedly a growing interest in genetic counselling, both in the medical profession and in the population generally. Requests for advice are becoming more numerous. Often family doctors or specialists can give advice. Sometimes reference to specialised clinics is desirable. So I should like to mention a brochure issued by the Advisory Medical Committee to the Ministry of Health to family doctors. It is called 'Medical Genetics' and was published in November 1967. A valuable feature is a list of clinics in England and Wales where specialised genetic advice is available. I should mention that these clinics are increasing in number and already the list is not quite complete. These brochures, however, are being constantly revised.

I have deliberately omitted some of the more general implications of genetic advice and what the effect may be on our reservoir of deleterious genes. In general, any change in frequencies is likely to be very slow. But this is a subject which will be dealt with fully by Dr Allison in the next article. He will also be dealing with the rapidly expanding field of abnormalities due to major errors of chromosome behaviour.

6 Genes and populations

Anthony Allison

You cannot study genetics in individuals; instead you have to study it in families or in populations. Family studies show which characters are inherited and the mode of inheritance – whether they are due to single genes or many genes, whether they are dominant or recessive, and so forth. But family studies do not tell us how common genes are: that is the province of population genetics. A population can be any group of breeding organisms. For example, the bacteria in my throat constitute a population, and so do all the field mice in the Isle of Wight or all the humans in the United Kingdom and Eire. The population geneticist determines how common particular genes are in different populations. He then tries to find out what factors influence the frequency of the genes in the populations – either tending to maintain the gene frequency constant or allowing it to change. Changes in gene frequency are important because that is what evolution is about.

As Darwin recognised, the first prerequisite for evolution is variation: you cannot have evolution if all forms of a particular organism are the same. Inherited variation is produced in the first place by mutation, which is due to small alterations in the genetic material, DNA. Usually, such mutations are disadvantageous and often the new forms or mutants are eliminated by natural selection. But occasionally the mutations are advantageous, and the mutant organisms and their descendants increase in numbers at the expense of the original type. Such evolutionary changes can occur very rapidly. The most dramatic are seen in single-celled organisms under strong selection, for example in bacteria exposed to drugs. Bacteria resistant to the drugs multiply and soon replace the original population. This can be a serious problem in chemotherapy, especially when there is linked resistance against several drugs, as is now happening with certain intestinal bacteria. Similarly, insects resistant to insecticides rapidly increase in numbers under selection and present serious problems to those who are trying to control insect-borne diseases in the tropics.

Now, many human mutations are disadvantageous, producing inherited disorders of one kind or another. Two well-known examples are haemophilia – the bleeding tendency which affected some of Queen Victoria's male descendants – and porphyria, an abnormality of blood pigments that produces sensitivity to light and sometimes disturbed mental function, as seems to have been the case with George III. There is some measure of selection against such genes, preventing their spread through human populations.

But other genes may be advantageous and increase in frequency. This is particularly obvious in the case of genes conferring resistance against disease. I have been interested in this problem for some years. Many genes are known in mice and other experimental animals to influence susceptibility to particular diseases. One example of commercial importance is the so-called Aleutian gene which produces a pale fur colour in mink. Coats made from the mutant mink are prized by ladies chiefly because they are more expensive than those from wild-type mink, and are therefore more effective as status symbols. Many mink breeders introduced the Aleutian gene into their animals. It was then found that the gene increased the susceptibility of mink to an infectious agent producing Aleutian disease. So it looks as though the scarcity value of the mutant mink will be maintained.

The most remarkable example of a human gene conferring resistance against disease is the sickle-cell gene which is common in many populations living in Central Africa, and also certain Mediterranean, near Eastern and Indian populations. In the presence of this gene, an abnormal haemoglobin pigment is formed in red blood cells. The gene is of special interest because it was the first one in which the chemical effect of a mutation was defined. In sickle-cell haemoglobin, one of the constituent amino-acids is different from that in normal haemoglobin. When sickle-cell haemoglobin loses oxygen it precipitates in cells and distorts them into sickle-shape. In all higher organisms genes exist in pairs like the animals in Noah's ark. Hence an individual can inherit one sickle-cell gene or two. If he has the misfortune to inherit two such genes this sickling change occurs in the circulating blood, so that anaemia and blockage of blood vessels follow. This produces a severe disease, and many affected children seldom survive to adulthood. Many thousands of African children die of this serious disease every year, and each death eliminates two sickle-cell genes from the population.

It can be shown that genes removed from the population so rapidly could not be replaced by new mutations. Why, then, has the gene remained so common? Over large parts of Central Africa the gene is found in 20 per cent of the population, and it rises to a maximum value of 40 per cent in some areas. The explanation is that individuals carrying one sickle-cell gene are resistant to infection by the parasite producing malignant tertian malaria, and so have a greater chance of surviving repeated infections in early childhood. Hence the gene has become common only in areas where malaria is endemic. This is an instructive example because hybrid individuals (those who possess one sickle-cell and one normal haemoglobin gene) have a selective advantage over pure-blooded individuals (who have either two sickle-cell genes or none at all). Through hybrid advantage a stable frequency of the gene – what is termed polymorphism – can be maintained indefinitely. But, if malaria were to be elim-inated, the frequency of the sickle-cell would be expected to fall; and there are indications that this is actually happening in Negro populations now living in non-malarious parts of the New World. This is not an isolated case, either. The absence of the enzyme, glucose-6-phosphate dehydrogenase, in red blood cells is another genetically-determined abnor-mality which is common in the tropics, and again appears to confer some measure of resistance against malaria. And in-dividuals of blood group A are known to be more liable to infection by influenza virus than individuals of other blood groups.

From such examples we can be quite sure that natural selection is operating in human populations, as it does in populations of lower animals. Natural selection is often thought of in terms of predation or fighting, but even in lower animals, resistance to disease is important and there is little doubt that disease represents the most potent agent of natural selection in man. This has interesting consequences. As infectious diseases come to be controlled by drugs and other measures, selection pressures on human populations are changing, and this may well alter frequencies of certain genes in human populations.

A medical success which may have such consequences is the elimination of disease due to incompatibility of Rhesus blood groups. This is another polymorphism, some individuals in most populations being Rh-positive and others Rh-negative. Some Rh-negative mothers bearing Rh-positive children become immunised against the Rh factor, and this can have ill effects on children subsequently borne. In severe cases, babies of sensitised mothers suffer brain damage or may even die. Such sensitisation has been quite common, and most of you will know of families in which it has occurred. Experiments and clinical trials carried out over several years by members of the Nuffield Genetics Unit in Liverpool and other centres have shown that administration of an Rh-antiserum under appropriate conditions can prevent immunisation of the mother, and it seems likely that serious disease due to Rh-incompatibility will be eliminated in a few years. Hence the selection which has operated against the Rh-negative gene will be relaxed, and it will be interesting to see whether this will affect the frequency of the Rh factors in Western European populations. The only drawback is that such a follow-up will require several generations – at least a hundred years – to produce meaningful results.

Genetic factors similar to those giving rise to blood groups also operate in tissue cells. These are the differences that give rise to rejection of transplanted organs such as skin, kidneys or hearts. Useful progress in typing of human tissues has been made during the past few years, and these should help to provide the closest possible match of donors and recipients before the operation. Such typing, along with the application of immunosuppressive procedures, should greatly increase the chances of successful transplantation of organs. This is an important aspect of biological individuality.

Another field of study producing important results is human population cytogenetics, that is, analysing how common abnormalities of the chromosome complement are in various populations and what kind of disorders they produce. Special attention has been given to the sex chromosomes because these are easily identified and because abnormalities are quite frequent. In a normal man or woman, there is one pair of sex chromosomes among the twenty-three pairs of chromosomes found in every cell in the body. In a woman the two sex chromosomes are alike and called X-chromosomes. A normal man has two dissimilar sex chromosomes, one X – the same as the female X – and a smaller chromosome called Y that determines maleness at conception.

About one fifth of one per cent of children are born with abnormal numbers of sex chromosomes, and this may impair normal sexual development and have other effects. Two of the commonest defects are known after the people who described them as Klinefelter's syndrome and Turner's syndrome. Men affected with Klinefelter's have certain female features, sometimes show breast development and are often mentally retarded. They are sterile because their testes are underdeveloped and fail to produce spermatozoa. These people have one or more extra X-chromosome, that is, their constitution is XXY, XXXY or, rarely, they may have four or more X-chromosomes. The probability of mental retardation increases with the number of X-chromosomes.

Women with Turner's syndrome are sterile, do not menstruate, are short of stature and often have congenital abnormalities such as a web of skin on either side of the neck or narrowing of the main artery after it leaves the heart. These women have only one X-chromosome instead of the usual two. There are numerous other varieties of intersex. They are fortunately rare in the population as a whole, but may be quite common in selected populations such as women athletes. It is now usual for women participating in international athletic contests to be examined by a panel of specialists, and several former record holders have been disbarred from further competition.

An important advance in knowledge has been made during the past few years by Dr Casey of Sheffield, Dr Jacobs of Edinburgh and their colleagues. In special hospitals for criminals, they found a number of men with an extra Y-chromosome. The incidence of this condition in the general population is not accurately known, but it must be far lower than in the criminal hospital population. The subjects with extra Y-chromosomes differed from others in the hospitals in several respects. The men with extra Y-chromosomes had no physical abnormalities which might account for their behaviour although they tended to be taller than average. They suffered from a severe disorder of personality, in most instances associated with some degree of intellectual impairment. Thus it seems that their intellectual function may have been insufficient to suppress the disordered drives leading to criminal behaviour. There is no reason to believe that these subjects would have indulged in crime had it not been for their abnormal personalities. There is no predisposing family environment, and their criminal activities often start at a remarkably early age before they are seriously influenced by factors from outside their homes. Moreover, repeated and prolonged attempts at corrective training have proved unsuccessful.

Thus it seems clear that in these males the inheritance of an extra Y-chromosome has resulted in a severely disordered personality, which has led these men into conflict with the law. This is a remarkable example in which a specific genetic abnormality can be associated with a tendency to criminal behaviour. It raises the question how far the tendency to crime in general is due to genetic factors, and how far it is due to unfavourable early environment. In either case there is a large measure of determination by factors beyond the control of the individual. This raises interesting moral issues and must limit the degree to which an individual can be blamed – or, for that matter, praised – for his behaviour in society. But the findings also make it clear that some people have such a strong predisposition to crime that corrective measures are unlikely to succeed. Some are dangerous in society; how to deal with the others is a difficult penal problem.

Cases of this sort are now coming before the courts. Richard Speck is said to be one: his attorneys are now preparing an appeal against his death sentence for killing eight nurses in Chicago in 1966. Another is Daniel Hugon, awaiting trial in Paris on a charge of having murdered a prostitute. His lawyers contend that he is mentally unfit to stand trial because of his chromosomal abnormality, and the Paris court has appointed a panel of experts, including a psychiatrist and cytogeneticist, to advise it.

Many other interesting facts about human chromosomes have been revealed during the last ten years. A number of rare chromosome defects have been associated with congenital disorders. But one is relatively common. This is the condition known as Down's syndrome or mongolism, in which the well-known physical appearance is associated with impairment of mental function and often with defects of the heart and other organs. Like the Klinefelter's syndrome, mongolism is due to an abnormal division of the eggs in the genital tract of the mother. This is more common in older mothers rising to a frequency of about one in two hundred and fifty children borne towards the end of reproductive life. Abnormal chromosome constitutions are also found in one third of naturally-occurring abortions. We do not know what is responsible for the chromosome aberrations in mongolism or in abortions. One possibility being actively pursued is that infections with viruses and other micro-organisms play a part. Such infections are known to affect chromosomes of cells in laboratory culture. If infections were responsible for some of these disorders, they might be preventable by immunisation or treatment. As it is, the best way of reducing the chances of having children with chromosomal abnormalities is for women to bear children before the age of thirty-five.

Children with mongolism are known to have an increased risk of developing leukaemia, a type of cancerous proliferation of white blood cells. The general relationship of chromosomal abnormalities and a tendency to develop cancer is also interesting. Subjects with one type of leukaemia regularly show loss of a small part of a chromosome. And children with certain congenital abnormalities known after the discoverers as Bloom's syndrome and Fanconi's anaemia, again show a tendency to develop leukaemia or skin cancer. When the cells from these children are grown in culture they show many chromosome breaks and other forms of abnormality. So many research workers believe, as I do, that chromosome mutations, caused by viruses, radiation or certain chemicals, are the basis of malignant proliferation of cells.

With the spectacular success in the treatment of many diseases during the past thirty years, the question arises what effects these will have on the frequencies of genes in human populations. Obviously, if diabetes has a hereditary component, and if diabetics live to have children, there is a possibility that the proportion of people carrying genes for susceptibility may increase. There is little doubt that this sort of effect is actually happening, although on a small scale which certainly does not justify eugenic intervention. But the opposite effect is also possible. In a population freed from malaria, for example, we can confidently predict that sickle-cell anaemia will gradually die out. Other effects cannot be predicted with certainty – for example effects resulting from reduced mortality due to the near elimination of infectious disease. Again, there has been much argument about effects on the general intelligence of the population due to a higher fertility of less intelligent people; but very little information is available. In any case, the spreading use of birth control in all social classes will probably change the trends during the next few decades.

7 Microbes and molecules

D. J. Cove

Viruses, bacteria and simple animals and plants are often referred to collectively by geneticists as micro-organisms. But micro-organisms do not represent a biologically closely-related group, they are united rather by the common factor of experimental convenience. Many of the experiments which have resulted in the considerable progress of our understanding of some of the most fundamental processes of life, have been performed using micro-organisms. But micro-organisms are not only convenient to work with, they are also of considerable importance in their own right. Not only do they include most of the disease-causing organisms, but also many of them produce substances of considerable economic importance. These range from antibiotics to alcohol, and include many raw materials important to industry. As yet little methodical breeding of micro-organisms has been carried out to increase their usefulness, but this should not prove difficult, and should lead to their becoming even more important to the chemical and pharmaceutical industries in the future.

But I do not wish to talk about the direct usefulness of micro-organisms now. Instead, I am going to speculate on the possible importance in the future of some of the fundamental discoveries made as a result of work on micro-organisms. Probably the most important of these discoveries were those which led to our understanding of the way in which hereditary information is used, and stored. The key to this process lies in the chemical deoxyribose nucleic acid, usually referred to as DNA.

DNA occurs in cells in the form of long molecules, which are made up of smaller sub-units, joined together in long strings. There are only four possible sub-units which can occur in DNA, and it now seems certain that the genetic information is determined by the order of these sub-units within the DNA molecule. So DNA is like a language which uses only four letters. But DNA on its own is of little use to the cell. The all-important molecules are proteins. Proteins have both structural and catalytic roles in cells, that is they are used to build the cell and also to guide chemical reactions within it. Those having a catalytic role are generally known as enzymes. Proteins, like DNA, are also long molecules built up of sub-units. There are however, at least twenty sub-units that can occur in proteins, and the order in which they are strung together is all important in deciding the properties of the whole protein molecule. But proteins themselves, are unable to serve as the hereditary material of the cell, firstly because the cell contains no means of replicating them, and secondly because they are too unstable. It is here in fact that DNA comes in. By a series of complex chemical reactions, the cell uses the sequence of the sub-units in the much more stable DNA molecules, to assemble the appropriate sub-units in the correct sequence to form specific proteins. We now believe that nearly all the properties of the cell are determined by the proteins it contains, and these in turn are determined by the cell's DNA, and so you will see how it is that DNA is able to serve as the cell's hereditary information. So it is probable that most of the improvements in animals and plants brought about by selective breeding, and also most inherited diseases, have their origin in alterations of the protein content of the cells, and therefore in the genetic information encoded in their DNA.

So it is quite reasonable to ask whether there is a possibility in the future of improving plants and animals, or eliminating hereditary diseases, not by the use of breeding programmes, but instead by injecting into the animal or plant the appropriate proteins, or even the hereditary information which directs the synthesis of these proteins. The proteins themselves do not appear to be very promising material for injection. As I mentioned earlier, proteins are generally unstable, and cannot be replicated. Also there is the problem of getting molecules such as proteins inside cells, and into the correct part of the cell. At first sight the hereditary information itself is equally unpromising. It is at present impossible to purify a specific part of the DNA content of an organism. Whereas a particular protein, an enzyme for example, can be recognised because of its catalytic activity, and can therefore be purified, one DNA molecule is chemically much the same as another.

It might eventually become possible to recognise a particular DNA molecule by devising a system which allowed it to direct the synthesis of the protein it coded for, but at the moment, we cannot do this and so we are unable to isolate the DNA specifying the structure of a particular protein. Even if we could purify this DNA, we would again, as with proteins, face the problem of getting the DNA into all the required cells of an organism. However, it is just possible that micro-organisms again might provide an answer as to how this could be done.

We are not alone in suffering from virus infection. Bacteria also suffer from parasitism by viruses. A virus particle consists of a protective protein coat, surrounding a core of nucleic acid, usually DNA, which serves as the virus's hereditary information. After attachment to the bacterium, the DNA alone enters the bacterial cell. Once in, the viral DNA begins at once to multiply, taking over the bacterial cell, and in the process of producing many progeny, killing it. But some viruses that infect bacterial cells are able to lead a curious double existence. Although after entry into the host bacterium they usually multiply and produce many progeny in the way I just mentioned, the viral DNA can instead sometimes become associated with the bacterial DNA in some way not yet fully understood. In this state, it does the bacterium no harm, but multiplies in step with the bacterial DNA so that all the progeny of the bacterium contain the viral DNA. Occasionally in some bacterial cells the viral DNA in the associated state may become dissociated from the bacterial DNA once again. Once dissociated it multiplies rapidly producing many virus particles and killing the host bacterium. In the process of dissociating from the bacterial DNA, the virus may, although only rarely, take with it a small part of the bacteria's genetic information, and when this happens the virus may leave behind some of its own. So it is possible to obtain viruses which carry an incomplete complement of their own hereditary information, but some of their host bacteria. Because they do not contain the complete viral information such viruses are often not virulent, but they can, under special conditions, be induced to multiply and so it is possible to obtain them in large quantities. By using an appropriate strain of such a virus to infect a bacterium deficient in its own information, we can repair this deficiency. Furthermore, by appropriate choice of virus we can ensure that it is unable to harm the bacterium.

So at least with bacteria we already know of a way in which a piece of genetic information can be injected into a large number of bacterial cells. We can now ask whether such a system might also be found so we could do the same thing to higher plants and animals, including man. At present we cannot be certain that any known virus parasitising higher organisms is able to associate with the DNA but there is some suggestive evidence that this may occur. If such a situation does exist, it might be possible firstly to grow donor cells in culture and then infect them with such a virus and so obtain quantities of virus carrying various pieces of hereditary information. These viruses could be used to infect, for example, a foetus while still in the uterus, if routine screening had detected that the foetus had a genetic deficiency. Such a procedure would have the advantage of avoiding rejection of the 'foreign' protein which the DNA specifies by the body's natural immunity mechanisms, which are not operative in the early foetus. Alternatively the seed of a crop plant might be infected with a virus, which carried the appropriate genetic information to confer disease resistance to the plant. Of course extreme care would have to be taken to ensure that the virus which acted as donor, did no harm to the recipient. The appropriate virus strains carrying donor DNA, and themselves doing no harm to their host might in fact be very difficult indeed to isolate. And so it seems worthwhile to speculate whether any other method could be adopted.

There is one alternative which might be used. Although five years ago it hardly seemed likely that we would be able to synthesise DNA with its sub-units in a defined sequence, already very considerable progress has been made in mastering the very complex techniques which are necessary to achieve such a task. I have already described the DNA molecule as being made up of many sub-units strung together in a defined order. In fact, each DNA molecule is made up of two such strings of sub-units wound round one another in a helix. You will recall that there are only four types of sub-unit in the DNA molecule.

These sub-units although related chemically, differ in their shape and precise structure. The double chain is such that in one chain, only one of the three remaining sub-units can lie opposite any particular sub-unit in the other chain. So the order of sub-units in each chain is mutually defined and complementary. This structure is of the utmost importance and accounts for the precision with which a DNA molecule can be replicated. If the two component chains of the DNA molecule unwind from around one another, they can each determine the synthesis of the complementary chain. In this way, two copies of the original molecule can be produced, and it is this process that occurs before cell division, so that each progeny cell can receive a complete set of hereditary information.

The fact that DNA has a double stranded structure, can also be exploited to simplify the task of synthesising DNA molecules of precise sequence. It is possible to produce single chains containing up to 20 sub-units by a chemical synthesis. If two such chains are synthesised so that 10 sub-units of one chain are exactly complementary to 10 of the other, the two chains will associate so that a structure will be formed which is 30 sub-units long overall, but made up of a double chained centre portion 10 sub-units long, with a

62

single chained region also 10 sub-units long at each end. Two such structures, providing the appropriate regions are complementary, can in turn associate to form a 50 sub-unit long structure still with single chained ends, but now with a 30 sub-unit long double section in the middle, and so on. Furthermore, enzymes are known which are able to join up the discontinuities in the chains which are formed when these partially single, partially double chained structures associate. By using these methods it should become possible to construct long defined sequences of DNA, without needing to synthesise any component larger than a single chain of 20 sub-units, and it may not even be necessary to synthesise single chains as long as this. I should stress at this stage, that this is a very new field, and it may be many years before it is possible to construct DNA molecules which contain the 1000 or so sub-units necessary to supply the information to specify the structure of a typical protein. But I think I would be being over-cautious if I said that such a synthesis would not be possible in the fairly near future.

Supposing then that it will be possible to synthesise DNA, and hence hereditary information to order, how might such a possibility be used to help man? It is true that by synthesis we could overcome the first of the two major problems which I suggested would make the task of supplying hereditary information to cells difficult. But it would still remain to get this hereditary information into the cells of the particular organism.

This might be done by again employing a virus. Viruses can after all be regarded as specialised structures for the injection of nucleic acid into cells and so it is not surprising that I should again think of making use of them for this purpose. The difficulty lies in devising a method of incorporating the synthetic hereditary information into the DNA complement of the virus. At least some of the viruses which parasitise bacteria possess DNA which at its end is only single chained. Were such a condition also found to apply to some of the viruses which could infect higher organisms, it might be possible to manufacture the synthetic DNA with a terminal single chained region, complementary to that at the end of the viral DNA. In this way we could contrive to stick the synthetic DNA and the viral DNA together in a similar way to that employed to build up the synthetic DNA itself. We would end up once again with a virus carrying donor DNA and we could use this to infect the deficient organism. I think personally that this second approach which I have described is likely to be the more promising in the future. I have so far chosen to ignore a difficulty which would arise in the case of plants. As far as I am aware all known plant viruses use as their hereditary information not deoxyribose nucleic acid, DNA, but ribose nucleic acid, RNA. It is unlikely that RNA viruses would associate stably with their host DNA and so it might not be possible to use viruses to transfer DNA into plants, and an alternative method might have to be sought.

So the day when it will be possible to add or correct abnormalities in the hereditary information of higher organisms may not be as far away as we would have thought a few years back, but in my opinion it will not be the product of the information, the protein, which will be donated, but the specific genetic material itself. I shall finally speculate, in what has been already a very speculative talk, on what uses we could put such a technique to, when and if it were perfected. I have already given as examples the correction of hereditary abnormalities in man, and this might initially be one of the most important uses. I also mentioned the possible use of the technique in order to transfer genetic information leading to disease resistance in plants and animals.

Eventually the genetic information which is necessary to raise yields in crop-plants, and other similar characteristics in animals might also be transferred, and finally the technique may also be of use to correct the informational abnormalities which seem likely to be the cause of many forms of cancer.

8 Man's choices

T. M. Sonneborn

The ancient and continuing effort to improve the quality of human life has taken on new features in modern times. It has become the intense concern of all of mankind, including the young, the less privileged, and minority groups. It has profited immensely by results of the physical and life sciences and technologies. And there is growing awareness that genetics could make further important contributions. Wide publicity has been given during the last few decades to each of the sensational advances of genetics and to their implications for man. Yet the public has had few opportunities to get a comprehensive view of what genetics has accomplished, what it can be expected to accomplish, and how these accomplishments might be used to improve the quality of human life. It was to supply this deficiency that the BBC produced this series on Genetic Engineering. Several of the previous contributors have had occasion to mention that genetic engineering raises conflicts with current standards of conduct. In this final contribution to the series, I have been asked to discuss these conflicts and the choices they raise.

The first choice that confronts us is whether or not to use human genetic engineering at all, in view of the fact that it always involves meddling, directly or indirectly, with human procreation. Such meddling is considered by many people to be fundamentally wrong, immoral and irreligious.

Actually, though, there are ways and ways of meddling with human procreation. Some of them are widely accepted and practised. Anything that influences choice of mates is meddling with human procreation. Parents do this when they arrange for their children to meet the right people and not the wrong people; or when they discourage, oppose, or prevent marriages outside the faith or race. In some societies, they actually choose the mate. This is not held to be morally objectionable, at least not on grounds of meddling with procreation; yet it clearly affects the choice of which couples shall join in reproduction. Moreover, as Professor Crow pointed out in the first talk, many economic, political and medical decisions profoundly affect not only the choice of mates but also the size of their families and the chances that their children will survive to reproduce. So these decisions, whether we are aware of it or not, very definitely meddle with human procreation and have genetic effects. While we may not approve of certain decisions, none of these forms of meddling is held to be immoral or irreligious. They are accepted as having social value.

So what forms of meddling are considered to be immoral or irreligious? To put the matter bluntly, it appears to be only the forms that meddle directly with sexual intercourse or that resort to procreation by other means. These are touchy and complicated problems; but the current verdict of mankind, judging by practice, is that even some of these forms of meddling are widely approved. Obviously, every form of birth control is meddling with sexual intercourse. Without going into the long history of its struggle for justification and acceptance, the fact is that birth control is now almost universally practised in large areas of the world. And in these regions, families belonging to those religions that *do not* officially sanction most forms of birth control use them, almost as much as families belonging to the religions that *do* sanction them. Public acceptance of birth control is tremendously important for our choices about genetic engineering, because a large part of its potential for the future, centres on birth control in one form or another.

The conflicts about birth control have shifted with time. At first they centred about the moral right to interpose, during intercourse, a block that prevented sperm from meeting egg, that is a block to fertilisation – the initiation of a new life. That issue was decided some time ago by the widespread practice of this form of birth control. The pill, working on a different principle, was then accepted as a matter of course. Currently, the question is about the moral right to prevent survival after fertilisation – the question of abortion. Opponents of abortion stand on the principle of the sanctity of every individual's life from the moment of conception – fertilisation; in other words, they hold that the earliest abortions are as much a form of murder as destroying imperfect babies after birth or for that matter, adults. On the other hand, proponents of abortion, pointing out that the earliest stages of development after fertilisation are formless, insensitive, and pre-human, hold that it is immoral to permit survival, with its inevitable misery of parents and child, when a foetus is destined to be grossly abnormal or defective.

While these discussions of principle go on, actions are being taken. The special case of abnormality resulting from drugs taken by the mother is now up for action in several countries and may set a precedent for more comprehensive later action. Abortion in other special cases, especially rape, has recently been legalised in several states of the USA. At least one widely-used method of birth control, the intrauterine device or coil, may actually work by inducing very early abortion, that is by permitting fertilisation to occur, but preventing implantation in the uterus, and so leading to the expulsion of a few-days-old embryo. When morning-after pills are developed, they will probably work in the same way. While we in the west appear to be moving more and more towards accepting the principle of abortion, the Japanese have already adopted it on a large scale as a preferred method of birth control.

What happened in past decades with the conflicts raised by contraceptives, and what is happening now with the conflicts raised by abortion, lead me to believe that present and future conflicts will be wisely resolved by enlightened public opinion formed on the basis of discussion and experience. The great numbers, heterogeneity and adventurous spirit of mankind assure that new possibilities will be tried by some people, regardless of how they stand in the light of current civil, moral, or sacred law. Some trials will fail to win general acceptance; others will succeed. Acceptance may at first be limited to special cases and later become more general and comprehensive. Submission to the test of public usage and opinion is a slow process; but this slow pace, permitting time for weighing and testing, affords a considerable measure of protection against unwise choices.

If, by this process, we eventually choose to sanction abortion, we could use it to eliminate certain kinds of genetic abnormality. Visible deviations from the normal set of chromosomes cause gross abnormalities, severe mental deficiency, and distortions of personality. As pointed out earlier in this series, it is now possible, without injury to mother or child, to obtain cells of foetal origin, but not part of the foetus, examine their chromosomes and, if they show certain deviations predict with assurance that the foetus will develop into a grossly abnormal child. We shall soon be faced with the choice of permitting or preventing abortion after such an adverse finding.

The same choice will have to be faced when it becomes generally possible, as it already is for a few genes, to detect in the foetus, single genes that have abnormal effects. The more difficult decision of how late in development abortion will be permitted cannot long be evaded because certain methods of detection of abnormality cannot yet be applied very early in development.

Different choices and moral problems are raised by other forms of meddling with reproduction, those that exploit techniques of reproduction without sexual intercourse – artificial fertilisation and artificial implantation. Sterility of the husband has long led many couples to have the wife artificially inseminated, usually by an anonymous donor of semen. Further, as Dr Edwards told you in his talk, methods may soon be developed that make it possible for certain sterile women to produce children. The methods would involve removing an egg from her, fertilising it and later, at an early stage of development, implanting it in her uterus. So far these methods have been tried only on animals. If they should become possible with human females, as they probably will, some women will surely try them. The uses of artificial insemination, egg removal and implantation could obviously be extended beyond cases of sterility; for example, to avoid transmitting undesired genes, but still have children.

These techniques raise no serious problems when egg and sperm are provided by wife and husband. When either comes from another person, legal and moral problems arise. The legal problems have long been solved, in principle, by sanctioning adoption of children produced by other donors of both egg and sperm, often out of wedlock. The legal aspect of the new technologies involves only one new feature, the earlier age of the child to be adopted; instead of adopting a fully-formed child, the couple would 'adopt' a fertilised egg in an early embryonic stage. And since the moral and religious problems concern production, not destruction, of a new life, they seem less formidable than those raised by abortion. If abortion wins acceptance, as now seems likely, the techniques of artificial insemination and artificial implantation will hardly be rejected in the long run.

What I have been saying up to now boils down to the simple contention that individuals and society have ways of sanctioning the uses of new knowledge and technology and that in fact they do when they want to, even if these uses initially conflict with current legal, moral and religious principles. If the principles are found to operate contrary to the present good of man, readjustments are eventually made. In other words, the touchstone of man's choices is simply 'is it good for man?'

What uses of genetic engineering will man choose as good? In my opinion, it is unrealistic to expect man's choices to be guided at present by long-range considerations of human evolution, at least not by enough people to have an appreciable effect. Relatively few people's reproductive choices are made with regard to anything beyond their own children. This is not to say, though, that future generations will always be as short-sighted.

Actually, that does not matter greatly. What does matter

is the choices made in each generation with regard only to the production of the next generation. Widespread wise choices, continued generation after generation, would add up to wise direction of human evolution. What choices is man making now ? What other choices could he make ?

He is certainly choosing to do as much as he can, by every possible means, to mitigate in himself the undesired effects of his genetic constitution. For example, he takes insulin to counteract genes for diabetes; he performs on babies tests for phenylketonuria and controls it by diet, when discovered early enough; he accepts organs from other people and will eventually accept them from animals genetically bred for the purpose. By such means, he reduces human misery now; but thereby he often permits the genes that cause it to survive and be passed on to later generations by individuals who would otherwise not survive to transmit these genes.

Man has also begun to take preventive genetic action, to prevent the conception or to abort the gestation of defective and abnormal children. People greatly desire not to have such children. Because this hits home hard, man will doubtless in time come to use for this purpose every available means offered by genetic and reproductive technology, including much more sophisticated techniques of genetic engineering which are already envisaged but not yet developed. And this is all to the good. It decreases human misery and raises the average quality of human life by decreasing the number of genetically defective people.

On the other hand, man has thus far chosen to do very little about taking constructive genetic action, action designed to produce, or increase the chance of producing, children not merely free from defect and abnormality, but by some standard above average. Will enough people want strongly enough to do this so that they will use whatever means are available to accomplish it ? What are the means ? What are the choices ?

The situation has been repeatedly set forth by my late colleague, H. J. Muller, a noted advocate of positive eugenics.

In essence, the principal present means is to make available to others the germ cells of individuals who have shown in high degree the qualities one wishes for in one's own children. This assumes, of course, that to some extent their genetic constitution was a factor – but not necessarily the only or even the major factor – in their superiority of performance. Muller emphasised taking advantage of the fact that sperm can be frozen, stored, then thawed and used long after the donor dies. When donation of eggs and artificial implantation becomes possible for humans as they are now for some animals, germ cells from both sexes could be used in this way to foster the production of above-normal children.

Obviously the legal, moral and religious problems are exactly the same as those discussed earlier in connection with using the same methods to overcome sterility or avoid abnormality. The choices then do not differ as to technological means, but only as to ends. If people come to desire to have above-average or outstanding children with anything like the strength of their desire to avoid having defective and abnormal children, they will find it as easy – or as difficult – to resolve the moral problem in the one case as in the other, for it is the same problem.

For some years, a relatively small number of people have, in fact, decided that this *is* good for man and they are trying it out. Muller believed that the results of the trials – the qualities of the children – would lead to wider and wider public acceptance and use. If such parental choices were made on a large scale, the quality of human life would indeed be raised little by little, generation after generation. If we do not so choose, we shall long be restricted largely to preventive technology, to the decrease of abnormality.

Does the future promise possibilities of constructive genetic choices by means other than parental choice? Indeed it does. Eventually, genes will be made to order, including genes not now possessed by man or even by any organism. Means of incorporating them into the germ plasm will be devised. In the same way, genes lacking in man, but present

71

in other animals or even plants, will be transplantable to man. This will enable man to acquire new functions and possess new powers. Even more; conceivably, man could eventually create a successor to himself that would be both simpler and superior to anything he could evolve from himself. These future possibilities are unforeseeably far in the future. I mention them only to emphasise that when we have made our choices, that will not be the end. New ones will keep coming up as far ahead as we can imagine.

But our present concern is with the here and now, with our present significant choices. My thesis has been that we are doing pretty well with our slow and cautious method of public discussion, individual decisions and actions, and public acceptance of what is considered good for man and rejection of what is considered bad. We agree that it is good to reduce misery and improve the quality of life, and that it is good to use both environmental and genetic means of achieving those objectives. We have shown and continue to show that our conceptions of what is right and moral can change in the light of new possibilities for improving the quality of life. What we lack is neither the flexibility of mind nor adventurous spirits, but knowledge and experience. If the future can be judged by the past and present, we shall get that knowledge and experience and eventually add to our repertoire of means of doing what is believed to be good for man.

What can you and I do about it now? Mainly keep open the lines of communication between the laboratory and the public, foster the studies and research on which further hope for genetic improvement depends, and join in the public discussion of new possibilities as they emerge and are tried out. I am confident of the outcome because I am confident that enough people will face up to each challenge and choice to keep issues lively until solutions satisfactory for the good of man are attained.

In essence, the principal present means is to make available to others the germ cells of individuals who have shown in high degree the qualities one wishes for in one's own children. This assumes, of course, that to some extent their genetic constitution was a factor – but not necessarily the only or even the major factor – in their superiority of performance. Muller emphasised taking advantage of the fact that sperm can be frozen, stored, then thawed and used long after the donor dies. When donation of eggs and artificial implantation becomes possible for humans as they are now for some animals, germ cells from both sexes could be used in this way to foster the production of above-normal children.

Obviously the legal, moral and religious problems are exactly the same as those discussed earlier in connection with using the same methods to overcome sterility or avoid abnormality. The choices then do not differ as to technological means, but only as to ends. If people come to desire to have above-average or outstanding children with anything like the strength of their desire to avoid having defective and abnormal children, they will find it as easy – or as difficult – to resolve the moral problem in the one case as in the other, for it is the same problem.

For some years, a relatively small number of people have, in fact, decided that this *is* good for man and they are trying it out. Muller believed that the results of the trials – the qualities of the children – would lead to wider and wider public acceptance and use. If such parental choices were made on a large scale, the quality of human life would indeed be raised little by little, generation after generation. If we do not so choose, we shall long be restricted largely to preventive technology, to the decrease of abnormality.

Does the future promise possibilities of constructive genetic choices by means other than parental choice? Indeed it does. Eventually, genes will be made to order, including genes not now possessed by man or even by any organism. Means of incorporating them into the germ plasm will be devised. In the same way, genes lacking in man, but present

in other animals or even plants, will be transplantable to man. This will enable man to acquire new functions and possess new powers. Even more; conceivably, man could eventually create a successor to himself that would be both simpler and superior to anything he could evolve from himself. These future possibilities are unforeseeably far in the future. I mention them only to emphasise that when we have made our choices, that will not be the end. New ones will keep coming up as far ahead as we can imagine.

But our present concern is with the here and now, with our present significant choices. My thesis has been that we are doing pretty well with our slow and cautious method of public discussion, individual decisions and actions, and public acceptance of what is considered good for man and rejection of what is considered bad. We agree that it is good to reduce misery and improve the quality of life, and that it is good to use both environmental and genetic means of achieving those objectives. We have shown and continue to show that our conceptions of what is right and moral can change in the light of new possibilities for improving the quality of life. What we lack is neither the flexibility of mind nor adventurous spirits, but knowledge and experience. If the future can be judged by the past and present, we shall get that knowledge and experience and eventually add to our repertoire of means of doing what is believed to be good for man.

What can you and I do about it now? Mainly keep open the lines of communication between the laboratory and the public, foster the studies and research on which further hope for genetic improvement depends, and join in the public discussion of new possibilities as they emerge and are tried out. I am confident of the outcome because I am confident that enough people will face up to each challenge and choice to keep issues lively until solutions satisfactory for the good of man are attained.

In essence, the principal present means is to make available to others the germ cells of individuals who have shown in high degree the qualities one wishes for in one's own children. This assumes, of course, that to some extent their genetic constitution was a factor – but not necessarily the only or even the major factor – in their superiority of performance. Muller emphasised taking advantage of the fact that sperm can be frozen, stored, then thawed and used long after the donor dies. When donation of eggs and artificial implantation becomes possible for humans as they are now for some animals, germ cells from both sexes could be used in this way to foster the production of above-normal children.

Obviously the legal, moral and religious problems are exactly the same as those discussed earlier in connection with using the same methods to overcome sterility or avoid abnormality. The choices then do not differ as to technological means, but only as to ends. If people come to desire to have above-average or outstanding children with anything like the strength of their desire to avoid having defective and abnormal children, they will find it as easy – or as difficult – to resolve the moral problem in the one case as in the other, for it is the same problem.

For some years, a relatively small number of people have, in fact, decided that this *is* good for man and they are trying it out. Muller believed that the results of the trials – the qualities of the children – would lead to wider and wider public acceptance and use. If such parental choices were made on a large scale, the quality of human life would indeed be raised little by little, generation after generation. If we do not so choose, we shall long be restricted largely to preventive technology, to the decrease of abnormality.

Does the future promise possibilities of constructive genetic choices by means other than parental choice ? Indeed it does. Eventually, genes will be made to order, including genes not now possessed by man or even by any organism. Means of incorporating them into the germ plasm will be devised. In the same way, genes lacking in man, but present

in other animals or even plants, will be transplantable to man. This will enable man to acquire new functions and possess new powers. Even more; conceivably, man could eventually create a successor to himself that would be both simpler and superior to anything he could evolve from himself. These future possibilities are unforeseeably far in the future. I mention them only to emphasise that when we have made our choices, that will not be the end. New ones will keep coming up as far ahead as we can imagine.

But our present concern is with the here and now, with our present significant choices. My thesis has been that we are doing pretty well with our slow and cautious method of public discussion, individual decisions and actions, and public acceptance of what is considered good for man and rejection of what is considered bad. We agree that it is good to reduce misery and improve the quality of life, and that it is good to use both environmental and genetic means of achieving those objectives. We have shown and continue to show that our conceptions of what is right and moral can change in the light of new possibilities for improving the quality of life. What we lack is neither the flexibility of mind nor adventurous spirits, but knowledge and experience. If the future can be judged by the past and present, we shall get that knowledge and experience and eventually add to our repertoire of means of doing what is believed to be good for man.

What can you and I do about it now? Mainly keep open the lines of communication between the laboratory and the public, foster the studies and research on which further hope for genetic improvement depends, and join in the public discussion of new possibilities as they emerge and are tried out. I am confident of the outcome because I am confident that enough people will face up to each challenge and choice to keep issues lively until solutions satisfactory for the good of man are attained.